SCHOLAR Study Guide

CfE Advanced Higher Physics
Unit 2: Quanta and Waves

Authored by:

Chad Harrison (Tynecastle High School)

Reviewed by:

Grant McAllister (Bell Baxter High School)

Previously authored by:

Andrew Tookey (Heriot-Watt University)

Campbell White (Tynecastle High School)

Heriot-Watt University

Edinburgh EH14 4AS, United Kingdom.

First published 2015 by Heriot-Watt University.

This edition published in 2015 by Heriot-Watt University SCHOLAR.

Copyright © 2015 SCHOLAR Forum.

Distributed by the SCHOLAR Forum.

SCHOLAR Study Guide Unit 2: CfE Advanced Higher Physics

1. CfE Advanced Higher Physics Course Code: C757 77

ISBN 978-1-911057-01-7

Print Production and fulfilment in UK by Print Trail www.printtrail.com

Acknowledgements

Thanks are due to the members of Heriot-Watt University's SCHOLAR team who planned and created these materials, and to the many colleagues who reviewed the content.

We would like to acknowledge the assistance of the education authorities, colleges, teachers and students who contributed to the SCHOLAR programme and who evaluated these materials.

Grateful acknowledgement is made for permission to use the following material in the SCHOLAR programme:

The Scottish Qualifications Authority for permission to use Past Papers assessments.

The Scottish Government for financial support.

The content of this Study Guide is aligned to the Scottish Qualifications Authority (SQA) curriculum.

Contents

Topic 1

Introduction to quantum theory

Contents

Prerequisite knowledge

- *Structure of atoms.*

- *Irradiance, photons and formation of line and emission spectra.*

- *The EM spectrum and properties of waves.*

- *Angular momentum.*

Learning objectives

By the end of this topic you should be able to:

- *state that the angular momentum of an electron about a nucleus is quantised in units of $h/2\pi$;*

- *state the equation $mvr = nh/2\pi$ and perform calculations using this equation;*

- *qualitatively describe the Bohr model of the atom;*

- *understand what is meant by black body radiation and the Ultra-Violet Catastrophe;*

- *state that quantum mechanics is used to provide a wider-ranging model of the atom than the Bohr model, and state that quantum mechanics can be used to determine probabilities;*

- *understand the Heisenberg Uncertainty Principle and be able to use the relationships*
$$\Delta x \Delta p \geqslant \frac{h}{4\pi}$$
$$\Delta E \Delta t \geqslant \frac{h}{4\pi}$$
.

1.1 Introduction

Towards the end of the 19th century, physical phenomena were described in terms of "classical" theory, as either particles or waves. However, some new discoveries (such as the photoelectric effect) could not be explained using classical theory. As we have seen, such phenomena required a theory that included a particle-like description of light. Analysis of Black Body Radiation curves showed us trends that could not be predicted by classical theory. We will see in this topic that a quantum approach was also used to answer questions about the structure of the atom. The hydrogen atom has the simplest structure, and we will look at a model of the hydrogen atom based on quantum theory and wave-particle duality. We will see that the emission spectrum of hydrogen is due to the electron moving between its allowed orbits, which can be determined by treating the electron as a wave.

Whilst the model of the atom dealt with here gives good agreement with experiments performed on the hydrogen atom, it is found to be unsuitable for larger atoms and molecules. The theory of quantum mechanics, which is used to describe such atoms, is introduced and we finish the topic by looking at Heisenberg's Uncertainty Principle.

1.2 Atomic models

The widely accepted view at the beginning of the 20th century was that an atom consisted of a large positively-charged mass with negatively-charged electrons embedded in it at random positions. This "plum pudding" model (suggested by J J Thomson) was consistent with experimental data available at the time. But the Rutherford scattering experiments first performed early in the 20th century were not consistent with that model, and required a major rethink. Rutherford interpreted the results of his experiments as evidence that the atom consisted of a relatively massive positively charged nucleus with electrons of far lower mass orbiting around it.

The study of atomic spectra also gave interesting results. The emission spectra of elements such as hydrogen, shown in Figure 1.1, consist of a series of discrete lines, rather than a continuous spectrum. In terms of photons, there was no explanation as to why a particular element would only emit photons with certain energies.

Figure 1.1: Emission spectrum of hydrogen

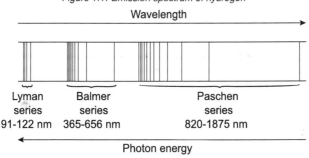

....

The next stage in the development of atomic theory was to try to build up a picture in which these experimental results could be explained.

1.2.1 The Bohr model of the hydrogen atom

In 1913 the Danish physicist Neils Bohr proposed an alternative model for the hydrogen atom. In the **Bohr model**, the electron orbits the nucleus (consisting of one proton) in a circular path, as shown in Figure 1.2. (This figure is not drawn to scale).

Figure 1.2: Simple model of the hydrogen atom

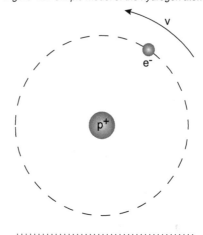

Bohr assumed that the electron could only have certain allowed values of energy, the total energy being made up of kinetic energy (due to its circular motion) and potential energy (due to the electrical field in which it was located). Each value of energy corresponded to a unique electron orbit, so the orbit radius could only take certain values. Using classical and quantum theory, Bohr calculated the allowed values of energy and radius, matching the gaps between electron energies to the photon energies observed in the line spectrum of hydrogen.

Since the electron is moving with speed v in a circle of radius r, its centripetal acceleration is equal to v^2/r. This presents a problem, as classical physics theory states that an accelerating charge emits electromagnetic radiation. A 'classical' electron would therefore be losing energy, and would spiral into the nucleus, rather than continue to move in a circular path. Bohr suggested that certain orbits in which the electron had an allowed value of angular momentum were stable. The angular momentum L of any particle of mass m moving with speed v in a circle of radius r is $L = mvr$. According to Bohr, so long as the angular momentum of the electron is a multiple of $h/2\pi$, the orbit is stable (h is Planck's constant). Thus the Bohr model proposed the concept of **quantisation of angular momentum** of the electron in a hydrogen atom. That is to say, the electron must obey the condition

$$\text{Angular momentum} = \frac{nh}{2\pi}$$

(1.1)

....

where n is an integer.

This quantisation of angular momentum fitted in with the predicted energy levels, but left a crucial question unanswered. Why were these particular orbits allowed? In other words, what made this value of angular momentum so special, and why did having angular momentum of $\frac{nh}{2\pi}$ make the orbit stable? The answer to this came when de Broglie's ideas on wave-particle duality were published a decade later. Treating the electron as a stationary wave, let us suppose the smallest allowed circumference of the electron's orbit corresponds to one wavelength λ. The next allowable orbit corresponds to 2λ, and so on. The orbit will be stable, then, if the circumference is equal to $n\lambda$. A stationary wave does not transmit energy, so if the electron is acting as a stationary wave then all its energy is confined within the atom, and the problem of a classical electron radiating energy does not occur.

Figure 1.3: Standing wave orbit of the electron in a hydrogen atom

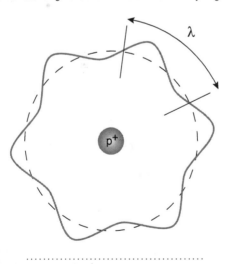

....

Figure 1.3 shows an electron in the $n = 6$ orbit. There are six de Broglie wavelengths in the electron orbit.
(De Broglie's work will be covered in more detail in Topic 2)

The circumference of a circle of radius r is $2\pi r$, so the condition for a stable orbit is

$$n\lambda = 2\pi r$$

(1.2)

. . . .

Rearranging the de Broglie equation (explained fully in Topic 2) $p = {}^{h}/_{\lambda}$

$$\lambda = \frac{h}{p} = \frac{h}{mv}$$

Now, we can substitute this expression for λ into Equation 1.2.

$$n\lambda = 2\pi r$$
$$\therefore \frac{nh}{mv} = 2\pi r$$
$$\therefore \frac{nh}{2\pi} = mvr$$
$$\therefore mvr = \frac{nh}{2\pi}$$

(1.3)

. . . .

Thus the angular momentum mvr is a multiple of ${}^{h}/_{2\pi}$, as predicted by Bohr (Equation 1.1). Treating the electron as a stationary wave gives us the quantisation of angular momentum predicted by Bohr. The unit of allowed angular momentum is sometimes given the symbol $\hbar$ ('h bar'), where $\hbar = {}^{h}/_{2\pi}$.

It should be pointed out that this result gives a good model for the hydrogen atom, but does not give us the complete picture. For instance, the electron is actually moving in three dimensions, whereas the Bohr atom only considers two dimensions. We will see later in this Topic that a full 3-dimensional wave function is used nowadays to describe an electron orbiting in an atom.

Example

In the $n = 2$ orbit of the hydrogen atom, the electron can be considered as a particle travelling with speed 1.09×10^6 m s^{-1}. Calculate:

1. the angular momentum of the electron;

2. the de Broglie wavelength of the electron;

3. the radius of the electron's orbit.

1. Since we have $n = 2$

$$L = \frac{nh}{2\pi}$$
$$\therefore L = \frac{2 \times 6.63 \times 10^{-34}}{2\pi}$$
$$\therefore L = 2.11 \times 10^{-34}\ \text{kg m}^2\,\text{s}^{-1}$$

2. The de Broglie wavelength is

$$\lambda = \frac{h}{p} = \frac{h}{mv}$$
$$\therefore \lambda = \frac{6.63 \times 10^{-34}}{9.11 \times 10^{-31} \times 1.09 \times 10^6}$$
$$\therefore \lambda = 6.68 \times 10^{-10}\ \text{m}$$

3. The angular momentum is quantised, so we can use Equation 1.3 to find r

$$mvr = \frac{nh}{2\pi}$$
$$\therefore r = \frac{2 \times 6.63 \times 10^{-34}}{2\pi mv}$$
$$\therefore r = \frac{1.326 \times 10^{-34}}{2\pi \times 9.11 \times 10^{-31} \times 1.09 \times 10^6}$$
$$\therefore r = 2.13 \times 10^{-10}\ \text{m}$$

. . . .

The Bohr model of the hydrogen atom

At this stage there is an online activity which allows you to view the electron as a particle or wave in the Bohr atom.

Go online

. . . .

1.2.2 Atomic spectra

The Bohr model tells us the allowed values of angular momentum for the electron in a hydrogen atom in terms of the quantum number n. For each value of n, the electron has a specific value of angular momentum L and total electron energy E. The Bohr model of the hydrogen atom allows E to be calculated for any value of n. When the electron moves between two orbits, it either absorbs or emits energy (in the form of a photon) in order to conserve energy, as shown in Figure 1.4.

Figure 1.4: (a) Emission and (b) absorption of a photon

photon
emitted

photon
absorbed

(a)   (b)

The line spectrum produced by atomic hydrogen allows us to calculate the difference
between energy levels. When an electron moves from a large n orbit to a lower n
orbit it loses energy, this energy being emitted in the form of a photon (Figure 1.4(a)).
Similarly, absorption of a photon raises the electron to a higher n orbit (Figure 1.4(b)).
The important point here is that only photons with the correct energy can be emitted or
absorbed. The photon energy must be exactly equal to the energy difference between
two allowed orbits. This is why the emission spectrum of hydrogen consists of a series
of lines rather than a continuous spectrum.

Hydrogen line spectrum

At this stage there is an online activity which explores how some of the lines in the visible
part of the hydrogen spectrum are produced.

Go online

Q1: If the electron makes a transition to a higher orbital e.g. $n = 2$ to $n = 3$ is a photon
absorbed or emitted?

Explain your answer.

Q2: Why does a transition from the $n = 1$ orbit produce Ultraviolet not visible light?

Quiz: Atomic models

Go online

Useful data:

Planck's constant h	6.63×10^{-34} J s
Mass of an electron m_e	9.11×10^{-31} kg

Q3: In Bohr's model of the hydrogen atom,

a) photons orbit the atom.
b) the electron's angular momentum is quantised.
c) the electron constantly emits electromagnetic radiation.
d) the electron constantly absorbs electromagnetic radiation.
e) the angular momentum is constantly changing.

....

Q4: What is the angular momentum of an electron orbiting a hydrogen atom, if the quantum number n of the electron is 4?

a) 1.67×10^{-32} kg m^2 s^{-1}
b) 8.33×10^{-33} kg m^2 s^{-1}
c) 8.44×10^{-34} kg m^2 s^{-1}
d) 4.22×10^{-34} kg m^2 s^{-1}
e) 2.64×10^{-35} kg m^2 s^{-1}

....

Q5: If an electron is orbiting a hydrogen atom with a de Broglie wavelength of λ and quantum number n, the radius r of the orbit is given by the equation

a) $r = 2\pi n\lambda$
b) $r = {n\lambda}/{h}$
c) $r = {2\pi}/{n\lambda}$
d) $r = {h}/{n\lambda}$
e) $r = {n\lambda}/{2\pi}$

....

Q6: The emission spectrum of hydrogen consists of a series of lines. This is because

a) photons can only be emitted with specific energies.
b) photons are re-absorbed by other hydrogen atoms.
c) the photons have zero momentum.
d) hydrogen atoms only have one electron.
e) once a photon is emitted it is immediately re-absorbed.

....

1.2.3 Black body radiation and the UV catastrophe

When objects are heated they can be thought of as **Black Bodies**, which radiate large amounts of energy in the form of infra-red radiation. Black Bodies have a cavity and a small hole to emit the infra-red radiation in a similar way to a point source of light. One example of a black body is a furnace. A small hole on the outside allows heat to enter and reflect off the inside before being absorbed by the walls of the furnace which then emit heat of a certain wavelength. The radiation reflects off the inside of the furnace and finally escapes and contains all frequencies of EM radiation with varying intensities.

Figure 1.5: A Furnace behaving like a black body

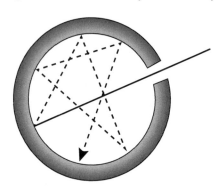

Stars are also approximate black body radiators. Stars absorb most incident light as well as other wavelengths of electromagnetic radiation and stars are able to emit all wavelengths of electromagnetic radiation too. Depending on the amount of heat the object will glow different colours ranging from cooler reds and oranges, through yellow and white up to a blueish white colour at very high temperatures.

It was discovered that there is a relationship between the amount of black-body radiation produced (more precisely called Specific Intensity when emitted by an object and Irradiance when incident on a surface) and the temperature of the black body. This graph shows the irradiance plotted against wavelength and it can be seen that at higher temperatures, the peak of the curve moves towards lower wavelengths or higher frequencies making the light appear closer to the blue end of the spectrum. Lord Rayleigh tried to use classical mechanics to devise an equation to solve this problem but this caused problems as at higher frequencies this predicted infinite energy emitted - this was called the Ultraviolet Catastrophe and was wasn't solved until 1900 when Planck produced a relationship which agreed with the black body curves found from experiments. Even Planck took many further years to believe the clear evidence for photons he'd discovered, Einstein's work on Photons (which he called "wave packets") in 1905 won him the Nobel prize and set the pattern for this exploration of the quantum world. Einstein's equations will be covered in more detail in Topic 2.

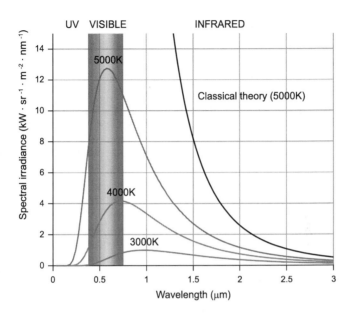

Black body radiation

There is an online activity allowing you to explore different predictions.

Go online

....

1.3 Quantum mechanics

In this section we will cover:

- Limitations of the Bohr hydrogen model
- Quantum mechanics

1.3.1 Limitations of the Bohr hydrogen model

The Bohr model gives an accurate prediction of the spectrum of atomic hydrogen. It also works well for hydrogen-like ions which have a single electron orbiting a nucleus, such as the helium (He^+) and lithium (Li^{2+}) ions. In these cases the larger central charge means that the electrons have different energies associated with the $n = 1,2,3...$ levels, but the underlying principle is the same - the electron has discrete energy levels, and the spectra of these ions are series of lines.

The model soon becomes inadequate when more electrons are added to the system. The motion of one electron is affected not only by the static electric field due to the nucleus, but also by the moving electric field due to the other electrons. The picture becomes even more complicated when we start dealing with molecules such as carbon dioxide. A more sophisticated theory is required to describe atoms and molecules with more than one electron.

1.3.2   Quantum mechanics

Nowadays **quantum mechanics** is used to describe atoms and electrons. In this theory the motion of an electron is described by a **wave function** Ψ. The idea of using a wave function was proposed by the German physicist Erwin Schrodinger, and Ψ is often referred to as the Schrodinger wave function. The motion is described in terms of probabilities, and the wave function is used to determine the probability of finding an electron at a particular location in three dimensional space. The electron cannot be thought of as a point object at a specific position; instead we can calculate the probability of finding the electron within a certain region, within a certain time period. Another German physicist, Werner Heisenberg, was also amongst the first to propose this concept as we will see later in the topic when looking at his Uncertainty Principle.

How does this relate to the Bohr hydrogen atom? The radii corresponding to different values of n predicted by Bohr are the positions of maximum probability given by quantum mechanics, so for the hydrogen atom the two theories are compatible. The allowed energy and angular momentum values calculated using the Bohr model are still correct. As we have mentioned already though, more complicated atoms and molecules require quantum mechanics.

1.4   Heisenberg uncertainty principle

The uncertainty principle was first developed in 1926 by Heisenberg who realised it was impossible to observe sub-atomic particles like electrons with a standard optical microscope no matter how powerful, because an electron has a diameter less than the wavelength of visible light. He came up with a thought experiment using a gamma ray microscope instead. Gamma rays with their higher frequency are much more energetic than visible light, so would change the speed and direction of the electron in an unpredictable way. Heisenberg then realised that in fact a standard optical microscope behaves similarly. To measure the position and velocity of a particle, a light can be shone on it, and then the reflection detected. On a macroscopic scale this method works fine, but on sub-atomic scales the photons of light that hit the sub-atomic particle will cause it to move significantly. So, although the position may have been measured accurately, the velocity or momentum of the particle is changed, and by finding out the position we lose any information we have about the particle's velocity. This can be said as "The very act of observation affects the observed."

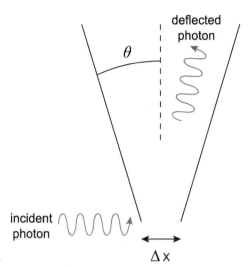

Heisenberg took this further and said that the values of certain pairs of variables cannot both be known exactly, so that the more precisely one variable is known, the less precisely the other can be known. An example of this is if the speed or momentum of a particle is known, then its location must be uncertain and vice versa. Similarly if we know the amount of energy a particle has, then we will not be able to determine how long it will have that energy for. In equation form this can be written as:

$$\Delta x \Delta p \geqslant \frac{h}{4\pi}$$

$$\Delta E \Delta t \geqslant \frac{h}{4\pi}$$

The minimum uncertainty in a particle's position **x** multiplied by the minimum uncertainty in a particle's momentum **p** has a minimum value about equal to the Planck constant divided by 4π. A similar relationship exists with Energy and time. Both can be shown to have the same units of Js.

This meant that particles could no longer be said to have separate, well-defined positions and velocities or energies and times, but they now exist in a "quantum state" which is a combination of their position and velocity or energy and time. You cannot know all the properties of the system at the same time, and any you do not know about in detail can be expressed in terms of probabilities.

A key question that needed answering about atoms was why do electrons orbiting the nucleus not lose energy and spiral inwards if they have a mass? Rutherford's model which had been accepted for the structure of the atom did not address this issue.

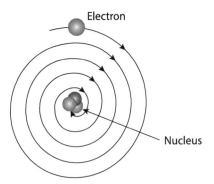

The answer is it is the uncertainty principle that prevents the electrons from getting too close to the nucleus. If they got too close we would precisely know their position and hence we would be very uncertain about their velocity. This uncertainty allows it to continue orbiting at the high speeds needed to stay in orbit. The uncertainty principle also helps explain why alpha particles can escape from the atom despite the strong force holding the nucleus together. The alpha particles use "quantum tunnelling" to escape (explained in more detail in Topic 2). Classical Physics says that they should not be able to escape.

Another example of this effect is nuclear fusion reactions taking place in the Sun. The Sun's temperatures are in actual fact not hot enough to give the protons the energy and speed they require to beat the massive repulsive electromagnetic force of the hydrogen atoms in the sun that they are trying to collide with. Thanks to quantum tunnelling and Heisenberg's uncertainty principle, the protons can apparently pass through the barrier despite the temperature and energy being approximately a thousand times too low for fusion to happen.

1.5   Extended information

Web links

There are web links available online exploring the subject further.

. . . .

1.6 Summary

> **Summary**
>
> You should now be able to:
>
> - state that the angular momentum of an electron about a nucleus is quantised in units of $h/2\pi$;
>
> - state the equation $mvr = nh/2\pi$ and perform calculations using this equation;
>
> - qualitatively describe the Bohr model of the atom;
>
> - understand what is meant by black body radiation and the Ultra-Violet Catastrophe;
>
> - state that quantum mechanics is used to provide a wider-ranging model of the atom than the Bohr model, and state that quantum mechanics can be used to determine probabilities;
>
> - understand the Heisenberg Uncertainty Principle and be able to use the relationships
> $$\Delta x \Delta p \geqslant \frac{h}{4\pi}$$
> $$\Delta E \Delta t \geqslant \frac{h}{4\pi}$$.

1.7 Assessment

End of topic 1 test

The following test contains questions covering the work from this topic.

Go online

Q7: Consider an electron orbiting in a Bohr hydrogen atom.

Calculate the angular momentum of the electron, if the quantum number n of the electron is 4.

_____ kg m^2 s^{-1}

. . . .

Q8: An electron in the $n = 1$ orbit of the hydrogen atom moves in a circle of radius 5.29 $\times$ 10^{-11} m around the nucleus.

Calculate the de Broglie wavelength of the electron.

_____ m

. . . .

Q9: An electron in a hydrogen-like atom in the $n = 3$ orbital has a de Broglie wavelength of 2.05 $\times$ 10^{-10} m.

Calculate the orbit radius of the electron.

_____ m

. . . .

An electron orbiting in the $n = 4$ level of a hydrogen-like atom drops to the $n = 2$ level, emitting a photon. The energy difference between the two levels is 3.19×10^{-19} J.

Q10: What is the energy of the emitted photon?

_____ J

. . . .

Q11: Calculate the wavelength of the emitted photon.

_____ m

. . . .

Q12: Which of the following statements are correct?

A) Perfect black bodies absorb and emit all wavelengths of radiation.

B) The ultraviolet catastrophe predicted energies tending to zero at higher frequencies of radiation emitted.

C) Black body radiation experiments showed higher frequencies of radiation emitted at higher temperatures.

D) Black Body radiation curves were more evidence that protons existed.

E) The Sun can be approximated to a black body.

. . . .

Q13: A bullet is travelling at a speed of 210 m s^{-1} and has a mass of 0.100 kg.

If the percentage uncertainty in the momentum of the bullet is 0.01%, what is the minimum uncertainty in the bullet's position if is measured at the exact same time as the speed is measured?

_____ m

. . . .

. . . .

Topic 2

Wave particle duality

Contents

Prerequisite knowledge

- *Wave particle duality, photons and a basic understanding of waves.*

- *Introduction to Quantum Theory (Topic 1).*

- *The Photoelectric Effect and Einstein's Energy Equation.*

Learning objectives

By the end of this topic you should be able to:

- *understand some examples of waves behaving as particles - the photo-electric effect and Compton Scattering;*

- *understand some examples of particles behaving as waves - electron diffraction and quantum tunnelling;*

- *be able to calculate the de Broglie wavelength using the equation $\lambda = {}^{h}/_{p}$ and to investigate the formation of standing waves around atoms and in wire loops.*

2.1   Introduction

Early in the 20th century, physicists realised that there were certain phenomena which could not be explained using the physical laws known at the time. Principally, phenomena involving light (thought to be composed of electromagnetic waves) could not be explained using this wave model. This topic begins with a look at some of the experiments that required a new theory - quantum theory - to explain their results. This theory explains that in some circumstances a beam of light can be considered to be made up of a stream of particles called photons. Light is said to exhibit "**wave-particle duality**",as both the wave model and the particle model can be used to describe the behaviour of light.

In the second part of the topic, this idea of wave-particle duality is explored further when it is shown that particles can sometimes exhibit behaviour associated with waves. Concentrating mainly on electrons, it will be shown that a stream of electrons can act in the same way as a beam of light passing through a narrow aperture. We finish the topic by looking at an interesting effect called quantum tunnelling which has many uses.

2.2   Wave-particle duality of waves

2.2.1   The photoelectric effect

The photoelectric effect was first observed in experiments carried out around 100 years ago. An early experiment carried out by Hallwachs showed that a negatively-charged insulated metal plate lost its charge when exposed to ultraviolet radiation. Other experiments, using equipment such as that shown in Figure 2.1, showed that electrons can be emitted from the surface of a metal plate when the plate is illuminated with light. This phenomenon is called the **photoelectric effect**.

Figure 2.1: Apparatus for photoelectric effect experiments

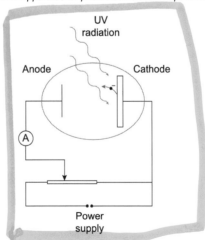

....

The equipment shown in Figure 2.1 can be used to observe the photoelectric effect. The cathode and anode are enclosed in a vacuum, with a quartz window to allow the cathode to be illuminated using an ultraviolet lamp (quartz is used because ordinary glass does not transmit ultraviolet light). A sensitive ammeter records the current in the circuit (the photocurrent). The potentiometer can be used to provide a "stopping potential" to reduce the photocurrent to zero. Under these conditions, the stopping potential provides a measure of the kinetic energy of the electrons emitted from the cathode (the **photoelectrons**), since when the photocurrent drops to zero, even the most energetic of the photoelectrons does not have sufficient energy to reach the anode.

In classical wave theory, the irradiance (power per unit area) of a beam of light is proportional to the square of the amplitude of the light waves. This means that the brighter a beam of light, the more energy is falling per unit area in any period of time, as you would expect. Yet experiments showed that the speed or kinetic energy of the emitted electrons did not depend on the irradiance of the beam. In fact the energy of the photoelectrons depended on the frequency (or wavelength) of the light. Below a certain frequency, no electrons were emitted at all, whatever the irradiance of the beam.

These observations cannot be explained using the wave theory of light. In 1905 Einstein proposed a quantum theory of light, in which a beam of light is considered to be a stream of particles ('quanta') of light, called **photons**, each with energy E, where

$$E = hf$$

(2.1)

....

In Equation 2.1, f is the frequency of the beam of light, and h is a constant, called Planck's constant after the German physicist Max Planck. The value of h is 6.63×10^{-34} J s. The equivalent expression in terms of the speed c and wavelength λ of the light waves is

$$E = \frac{hc}{\lambda}$$

Example

Calculate the energy of a photon in the beam of light emitted by a helium-neon laser, with wavelength 633 nm.

The frequency of the photon is given by the equation

$$f = \frac{c}{\lambda}$$
$$\therefore f = \frac{3.00 \times 10^8}{6.33 \times 10^{-7}}$$
$$\therefore f = 4.74 \times 10^{14} \text{ Hz}$$

The photon energy E is therefore

$$E = hf = 6.63 \times 10^{-34} \times 4.74 \times 10^{14}$$
$$\therefore E = 3.14 \times 10^{-19} \text{ J}$$

. . . .

The quantum theory explains why the kinetic energy of the photoelectrons depends on the frequency, not the irradiance of the incident radiation. It also explains why there is a 'cut-off frequency', below which no electrons are emitted.

When a photon is absorbed by the cathode, its energy is used in exciting an electron. A photoelectron is emitted when that energy is sufficient for an electron to escape from an atom. The cut-off frequency corresponds to the lowest amount of energy required for an electron to overcome the attractive electrical force and escape from the atom. The conservation of energy relationship for the photoelectric effect is given by Einstein's photoelectric equation

$$hf = hf_0 + \tfrac{1}{2}m_e v_e^2$$

(2.2)

. . . .

In Equation 2.2, hf is the energy of the incident photon and $\frac{1}{2}m_e v_e^2$ is the kinetic energy of the photoelectron. hf_0 is called the **work function** of the material, and is the minimum photon energy required to produce a photoelectron.

Example

Monochromatic ultraviolet radiation of frequency 1.06×10^{15} Hz is focused onto a magnesium cathode in a photoelectric effect experiment. The kinetic energy of the photoelectrons produced is 1.13×10^{-19} J. What is the work function of magnesium?

The photon energy is

$$E = hf = 6.63 \times 10^{-34} \times 1.06 \times 10^{15}$$
$$\therefore E = 7.03 \times 10^{-19} \text{ J}$$

Rearranging Einstein's equation, the work function is

$$hf_0 = hf - \tfrac{1}{2}m_e v_e^2$$
$$\therefore hf_0 = (7.03 \times 10^{-19}) - (1.13 \times 10^{-19})$$
$$\therefore hf_0 = 5.90 \times 10^{-19}\,\text{J}$$

....

2.2.2 Compton scattering

Experiments carried out in 1923 by the American physicist Arthur Compton provided further evidence to support the photon theory. He studied the scattering of a beam of X-rays fired at a thin sheet of graphite and found that some of the emergent beam was scattered into a wide arc. Furthermore the scattered X-rays had a slightly longer wavelength than the incident beam.

Compton explained his observations in terms of a collision between a photon and an electron in the graphite sheet, as shown in Figure 2.2.

Figure 2.2: Compton scattering from the collision of a photon with an electron

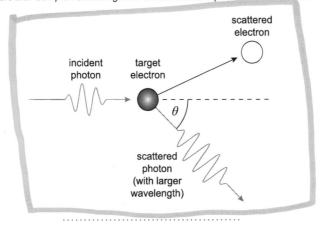

....

The incident X-rays have frequency f, so an individual photon has energy $E = hf$.

It was found that the scattered photons had frequency f', where $f' < f$.

We can compare the Compton scattering experiment to a moving snooker ball colliding with a stationary snooker ball. The X-ray is acting like a particle colliding with another particle, rather than a wave. Assuming the electron is initially at rest, the equation of conservation of energy is

$$hf = hf' + E_{k\ of\ recoil\ electron}$$

The conservation of energy relationship shows that $f' < f$, since the recoil energy of the electron $E_{k\ recoil}$ clearly has to be greater than zero as it is moving away from the collision. The wavelength of the scattered photon is $\lambda' = {}^c/_{f'}$, so the scattered wavelength is larger than the incident wavelength.

Compton performed experiments that confirmed this wavelength shift, which cannot be explained using a wave model of light. **Compton scattering** is therefore another example of the particle nature of light.

Compton also used the conservation of linear momentum in his analysis. Momentum is a property associated with particles, rather than waves, so Compton scattering is further proof of the dual nature of light. Using Einstein's energy-mass relationship from special relativity, the energy of a particle is related to its mass by the equation

$$E = mc^2$$

We can also express the energy of a photon in terms of its frequency f or wavelength λ

$$E = hf = \frac{hc}{\lambda}$$

Combining these two equations

$$mc^2 = \frac{hc}{\lambda}$$
$$\therefore mc = \frac{h}{\lambda}$$
$$\therefore p = \frac{h}{\lambda}$$

(2.3)

. . . .

The momentum p of a photon is given by Equation 2.3.

Equation 2.3 is a particularly important equation, as it demonstrates the wave-particle duality of light. The equation relates one property associated with waves, the wavelength λ, to a property associated with particles, namely the momentum p.

Quiz: Wave-particle duality of waves

  Useful data:

Go online

Planck's constant h	6.63×10^{-34} J s
Speed of light c	3.00×10^8 m s^{-1}

Q1: An argon laser produces monochromatic light with wavelength 512 nm. What is the energy of a single photon emitted by an argon laser?

a) 3.88×10^{-19} J
b) 1.29×10^{-27} J
c) 1.02×10^{-31} J
d) 3.39×10^{-40} J
e) 1.13×10^{-48} J

....

Q2: The work function of aluminium is 6.53×10^{-19} J. What is the minimum frequency of light which can be shone on a piece of aluminium to produce photoelectrons?

a) 1.01×10^{-15} Hz
b) 9.85×10^{14} Hz
c) 2.95×10^{23} Hz
d) Cannot say without knowing the wavelength of the light.
e) Cannot say without knowing the kinetic energy of the electrons.

....

Q3: In a Compton scattering experiment,

a) the momentum of the photons is unchanged after scattering.
b) the kinetic energy of the photons is unchanged after scattering.
c) the wavelength of the scattered photons decreases.
d) the wavelength of the scattered photons increases.
e) the wavelength of the scattered photons is unchanged.

....

Q4: A source of light contains blue (λ = 440 nm) and red (λ = 650 nm) light. Which *one* of the following statements is true?

a) The blue photons have greater photon energy and greater photon momentum.
b) The blue photons have greater photon energy, but the red photons have greater photon momentum.
c) The red photons have greater photon energy, but the blue photons have greater photon momentum.
d) The red photons have greater photon energy and greater photon momentum.
e) All photons have the same photon energy.

....

Q5:   Calculate the momentum of a single microwave photon emitted with wavelength 25.0 mm.

a) 7.96×10^{-24} kg m s^{-1}
b) 7.96×10^{-27} kg m s^{-1}
c) 2.65×10^{-32} kg m s^{-1}
d) 2.65×10^{-35} kg m s^{-1}
e) 5.53×10^{-44} kg m s^{-1}

. . . .

2.3   Wave-particle duality of particles

We have seen that light can be described in terms of a beam of particles (photons) rather than as transverse waves, under certain circumstances. The equation $\lambda = {}^h\!/_p$ (Equation 2.3) relates a property of waves (the wavelength λ) to a property of particles (the momentum p).

In 1924, Louis de Broglie (pronounced 'de Broy') proposed that this equation could also be applied to particles. That is to say, he suggested that under certain circumstances particles would behave as if they were waves, with a wavelength given by the above equation. Within three years, experiments with electron beams had proved that this was indeed the case. The wavelength given by $\lambda = {}^h\!/_p$ for a particle is now known as the **de Broglie wavelength**. As the following example shows, in most everyday cases the de Broglie wavelength is extremely small.

Example

Find the de Broglie wavelength of:

1. an electron (m_e = 9.11 × 10^{-31} kg) travelling at 4.00 × 10^5 m s^{-1};

2. a golf ball (m_b = 0.120 kg) travelling at 20 m s^{-1}.

1. The momentum p_e of the electron is

$$p_e = m_e v_e = 9.11 \times 10^{-31} \times 4.00 \times 10^5 = 3.64 \times 10^{-25} \text{ kg m s}^{-1}$$

The de Broglie wavelength of the electron is therefore

$$\lambda_e = \frac{h}{p_e} = \frac{6.63 \times 10^{-34}}{3.64 \times 10^{-25}} = 1.82 \times 10^{-9} \text{ m}$$

2. The momentum p_b of the golf ball is

$$p_b = m_b v_b = 0.120 \times 20 = 2.40 \,\mathrm{kg\,m\,s^{-1}}$$

The de Broglie wavelength of the golf ball is therefore

$$\lambda_b = \frac{h}{p_b} = \frac{6.63 \times 10^{-34}}{2.40} = 2.76 \times 10^{-34} \,\mathrm{m}$$

The de Broglie wavelength we have calculated for the golf ball is extremely small, so small that we do not observe any wave-like behaviour for such an object.

. . . .

Whilst Equation 2.3 can be used to calculate the de Broglie wavelength, it should be noted that if a particle is travelling at greater than about $0.1\ c$, the momentum must be determined using a relativistic calculation. Such a calculation is beyond the scope of this course.

De Broglie also suggested that electrons have orbits similar to standing waves and the electron waves form into integer number of wavelengths around the circumference as below:

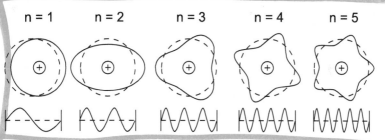

This can be demonstrated nicely by using a metal wire loop or even a metal "Chladni" plate on top of a vibrating object. A sequence of standing waves (see topic 5 for an explanation of standing waves) will be seen around the loop similar to electron standing waves around an atom.

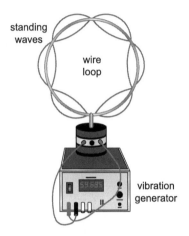

2.3.1   Diffraction

Since we can now calculate the wavelength associated with a moving particle, under what circumstances can we expect a particle to exhibit wave-like behaviour?

There are several aspects of the behaviour of light which can only be explained using a wave model. Interference, for example, occurs when two waves overlap in time and space, and the irradiance of the resultant wave depends on whether they interfere constructively or destructively.

Figure 2.3: (a) Constructive and (b) destructive interference of two sine waves

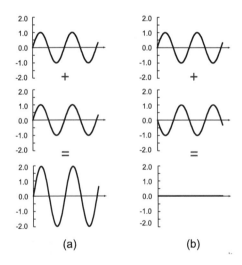

Constructive interference (Figure 2.3(a)) of two identical light waves would result in very bright illumination at the place where they overlap. Destructive interference (Figure 2.3(b)) results in darkness - the two waves "cancel each other out". Such an effect cannot usually be seen with particles - if two people kick footballs at you, you will feel the effects of both of them hitting you - they won't cancel each other out!

Another property of waves, which particles do not usually exhibit is **diffraction**. If a wave passes through an aperture, the size of which is about the same as the wavelength of the light, the light "spreads out" as it emerges from the aperture. This effect is shown in Figure 2.4, where the wavefronts of two sets of waves approaching two different apertures are shown.

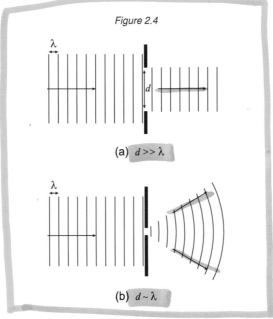

Figure 2.4

(a) $d >> \lambda$

(b) $d \sim \lambda$

In Figure 2.4(a), the diameter d of the aperture is much greater than the wavelength λ, and the waves incident on the aperture pass through almost unaffected. In Figure 2.4(b), the waves have a wavelength that is of the same order of magnitude as the aperture diameter, so the waves are diffracted and spread out as they emerge from the aperture.

An analogy for particles would be a train travelling through a tunnel. The width of the tunnel is only slightly larger than the width of the train, so one might expect to observe diffraction. Of course this doesn't happen - the train passes through the tunnel and continues travelling along the track. We don't observe diffraction because the de Broglie wavelength associated with the train is many orders of magnitude smaller than the tunnel width, as we have already seen with the example of the golf ball.

Our earlier example showed that for electrons, the de Broglie wavelength is much larger, although still a very small number. But this wavelength ($\sim 10^{-9}$ to 10^{-10} m) is about the same as the distance between atoms in a crystalline solid. The first proof of de Broglie's hypothesis came in 1927 when the first **electron diffraction** experiments were performed independently by Davisson and Germer in the United States and G. P. Thomson in Aberdeen.

Figure 2.5: Electron diffraction (a) schematic diagram of the experiment; (b) diffraction pattern recorded on the photographic plate

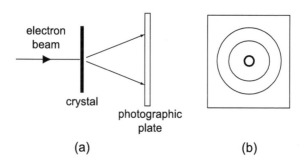

The spacing between atoms in a solid is typically around 10^{-10} m, so an electron travelling with a suitable momentum between two atoms in a crystal could be diffracted by the atoms. In a crystalline solid the atoms are arranged in a regular pattern or array, and act like a diffraction grating. Strong diffraction occurs in specific directions, which are determined by the atomic spacing within the crystal and the wavelength of the incident beam. Figure 2.5 shows the experimental arrangement and a typical diffraction pattern obtained. If the wavelength of the beam is known, the diffraction pattern can be used to determine the atomic spacing and crystal structure.

2.3.2   The electron microscope

In an optical microscope a sample is illuminated with white light and viewed through a series of lenses. The resolving power, or resolution, of the microscope is limited by the wavelength of the light. A microscope allows us to view a magnified image of very small objects, or view the fine structure of a sample. The resolving power tells us the smallest particles that can be distinguished when viewing through a microscope. The resolution is diffraction-limited, because light will be spread out when passing between two objects placed about one wavelength apart, making the two objects indistinguishable.

An electron microscope has a much smaller wavelength than visible light. Typically, the de Broglie wavelength of electrons in an electron microscope is around 10^{-10} m whilst the wavelength of visible light is around 10^{-7} m. Using an electron beam instead of a light source to 'illuminate' the sample means smaller features of the sample can be distinguished.

Quiz: Wave-particle duality of particles

Go online

Useful data:

Planck's constant h	$6.63 \times 10^{-34}\ J\ s$
Speed of light c	$3.00 \times 10^8\ m\ s^{-1}$

Q6: de Broglie's theory suggests that

a) particles exhibit wave-like properties.
b) particles have momentum and kinetic energy.
c) only waves can be diffracted.
d) photons always have zero momentum.
e) photons and electrons have the same mass.

....

Q7: What is the de Broglie wavelength of an electron (mass $m_e = 9.11 \times 10^{-31}$ kg) travelling with velocity 6.40×10^6 m s^{-1} ?

a) 1.75×10^{-15} m
b) 2.43×10^{-12} m
c) 2.47×10^{-12} m
d) 1.14×10^{-10} m
e) 1.56×10^{-7} m

....

Q8: A particle has a speed of 3.0×10^6 m s^{-1}. Its de Broglie wavelength is 1.3×10^{-13} m. What is the mass of the particle?

a) 1.5×10^{-38}kg
b) 7.8×10^{-32}kg
c) 1.7×10^{-27}kg
d) 5.1×10^{-23}kg
e) 5.1×10^{-21}kg

....

Q9: The demonstrations of electron diffraction by Davisson & Germer and G P Thomson showed that

a) electrons have momentum.
b) particles can exhibit wave-like properties.
c) light cannot be diffracted.
d) electrons have zero kinetic energy.
e) electrons have zero rest mass.

....

Q10: An electron microscope can provide higher resolution than an optical microscope because

a) electrons cannot undergo diffraction.
b) the electrons have a shorter wavelength than visible light.
c) electrons are particles whereas light is waves.
d) photons cannot be focused properly.
e) photons have a smaller momentum than electrons.

....

2.4   Quantum tunnelling

One very strange aspect of quantum physics is Quantum Tunnelling. This occurs due to the uncertainty of the position of objects such as electrons. There is a finite probability that an object trapped behind a barrier without enough energy to overcome the barrier may at times appear on the other side of the barrier. This process actually has no effect on the barrier itself - it is not broken down by this action. One example is electrons approaching the nucleus of atoms - there is some probability, no matter how small, that the electron will appear on the other side of the electromagnetic field which repels it.

This occurrence of appearing the other side of the nuclei is known as quantum tunnelling, and is easiest to visualise by thinking of the electron as a wave rather than a particle - a particle clearly cannot pass through the barrier but a wave may slightly overlap the barrier and even though most of the wave is on one side the small part of the wave that does cross the barrier leads to a small chance of the particle that generated the wave appearing on the far side of the barrier. This is shown in the following diagram and the video links below:

Quantum tunnelling through a barrier

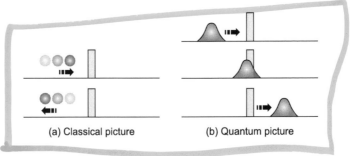

(a) Classical picture   (b) Quantum picture

Note the classical physics particle is repelled by the electromagnetic force, but the quantum physics wave has a small probability of tunnelling through the barrier.

This technique can be used to make quantum tunnelling composites (QTCs) which are materials which can be used as highly sensitive pressure sensors (due to the ability of quantum tunnelling to cross thin barriers and in electronic devices such as modern

Tunnel FET or TFET transistors replacing the older MOSFET transistors. Possible uses of the pressure sensors would include measuring the forces of impacts e.g. on fencing helmet or boxing glove or to accurately read a person's blood pressure.

2.5 Extended information

Web links

There are web links available online exploring the subject further.

. . . .

2.6 Summary

┌─ **Summary** ───
│ You should now be able to:
│
│ • understand some examples of waves behaving as particles - the photo-
│   electric effect and Compton Scattering;
│
│ • understand some examples of particles behaving as waves - electron
│   diffraction and quantum tunnelling;
│
│ • be able to calculate the de Broglie wavelength using the equation $\lambda = \frac{h}{p}$
│   and to investigate the formation of standing waves around atoms and in wire
│   loops.
└──

2.7 Assessment

End of topic 2 test

The following test contains questions covering the work from this topic.

Go online

Q11: Photons emitted by a monochromatic light source have energy 3.75×10^{-19} J.
Calculate the frequency of the light.

$f =$ _____ Hz

. . . .

Q12: Photons in a beam of monochromatic infrared radiation each have individual photon energy 5.35×10^{-20} J.
What is the wavelength of the infrared beam?

$\lambda =$ _____ m

. . . .

Q13: Light with a maximum frequency 8.35×10^{14} Hz is shone on a piece of metal. The work function of the metal is 3.65×10^{-19} J.

Calculate the maximum kinetic energy of the electrons produced due to the photoelectric effect.

$E_k =$ _____ J

. . . .

Q14: A laser emits monochromatic light with wavelength 425 nm.

Calculate the momentum of a single photon emitted by the laser.

$p =$ _____ kg m s^{-1}

. . . .

Q15: A filament lamp emits light with a wavelength range of 300 to 725 nm.

Calculate the largest value of photon momentum of the photons emitted by the lamp.

$p =$ _____ kg m s^{-1}

. . . .

A proton and an electron are both travelling at the same velocity.

Q16: Which particle has the greater momentum?

a) The proton
b) The electron
c) Neither - both have the same momentum

. . . .

Q17: Which particle has the greater de Broglie wavelength?

a) The proton
b) The electron
c) Neither - both have the same de Broglie wavelength

. . . .

In a television tube, electrons are accelerated by a high voltage and strike the screen with an average velocity of 6.35×10^6 m s^{-1}.

Q18: On average, what is the momentum of an electron just before it strikes the screen?

$p =$ _____ kg m s^{-1}

. . . .

Q19: Calculate the average de Broglie wavelength of the electrons.

$\lambda_e =$ _____ m

. . . .

Q20:

Complete the paragraph using some of the following words.

uncertainty	wave	electron
fission	neutron	proton
fusion	barrier	transistors

Quantum tunnelling is when an incident _____ is thought of as a _____ and part of the wave crosses a _____ . Due to the _____ of the wave's position, this allows the electron to pass through the barrier. This is how nuclear _____ is possible in the sun and how modern _____ have become so efficient and tiny.

. . . .

. . . .

Topic 3

Magnetic fields and particles from space

Contents

Prerequisite knowledge

- *Circular Motion.*

Learning objectives

By the end of this topic you should be able to:

- *explain what is meant by magnetic field and magnetic force;*

- *explain the behaviour of charged particles in a magnetic field;*

- *have an understanding of cosmic rays and the solar wind and how they interact with the Earth's atmosphere and magnetic field.*

3.1   Introduction

We are all familiar with permanent magnets and the fact that they exert forces on each other, as well as on certain types of metal and metallic ores. The first descriptions of magnetic effects were made in terms of **magnetic poles**. Every magnet has two poles. Unlike or opposite magnetic poles exert forces of attraction on each other, while like or similar poles repel each other. In addition, both poles of a magnet exert forces of attraction on unmagnetised iron. One pole is called the north pole or N-pole (actually short for "north-seeking" pole). It points approximately towards the north geographic pole of the Earth. The other end is called the south pole (S-pole). This alignment happens because the Earth is itself a magnet, with a south magnetic pole near to the north geographic pole.

It is tempting to compare north and south magnetic poles with positive and negative charges. While there are similarities, the major difference is that it is possible for isolated positive and negative charges to exist but there is no evidence to suggest that an isolated magnetic pole (a monopole) can exist.

In this topic, we will start by defining what is meant by a magnetic field and magnetic induction. We will then be looking at what happens when charged particles enter magnetic fields at various angles and will finally look at cosmic rays and the solar wind and how they interact with the Earth's atmosphere and magnetic field.

It is usual to describe the interaction between two magnets by applying a field description to them. We can consider that one magnet sets up a magnetic field and that the other magnet is in this magnetic field. A magnetic field is usually visualised by drawing magnetic field lines. Figure 3.1 shows the field pattern around a bar magnet.

Figure 3.1: The field pattern around a bar magnet

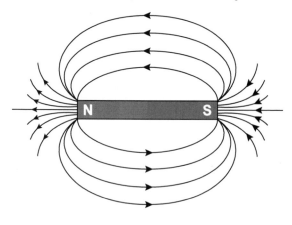

....

The following points should be noted about magnetic field lines:

- they show the direction in which a compass needle would point at any position in the field;

- they never cross over each other because the direction of the magnetic field is unique at all points - crossing field lines would mean that the magnetic field pointed in more than one direction at the same place - obviously impossible;

- they are three-dimensional, although this is not often seen on a page or screen - because the magnetic field they represent is three-dimensional;

- they indicate the magnitude of the magnetic field at any point - the closer the lines are together, the stronger is the field.

- like electric field lines, magnetic field lines are used to visualise the magnitude and the direction of the field. Also like electric field lines, they do not exist in reality.

An atom consists of a nucleus surrounded by moving electrons. Since the electrons are charged and moving, they create a magnetic field in the space around them. Some atoms have magnetic fields associated with them and behave like magnets. Iron, nickel and cobalt belong to a class of materials that are **ferromagnetic**. In these materials, the magnetic fields of atoms line up in regions called **magnetic domains**. If the magnetic domains in a piece of ferromagnetic material are arranged so that most of their magnetic fields point the same way, then the material is said to be a magnet and it will have a detectable magnetic field.

The Earth's magnetic field is thought to be caused by currents in the molten core of the Earth. A simplified view of the Earth's magnetic field is that it is similar to the field of a bar magnet. This means that the field lines are not truly horizontal at most places on the surface of the Earth, the angle to the horizontal being known as the magnetic inclination. The Earth's magnetic field at the poles is vertical. A compass needle is simply a freely-suspended magnet, so it will point in the direction of the Earth's magnetic field at any point. The fact that the magnetic and geographic poles do not exactly coincide causes a compass needle reading to deviate from geographic north by a small amount that depends on the position on the Earth. This difference is known as the magnetic declination or the magnetic variation.

3.2 Force on a moving charge in a magnetic field

$$F = BQ \, v \sin \theta$$

(3.1)

. . . .

Where F is the magnetic force experienced by a charged particle in Newtons, B is the magnetic field strength measured in Tesla, Q is the size of the charged particle in Coulombs and θ is the angle between B and v. The component of v that is perpendicular to the magnetic field is $v \sin \theta$.

To work out the direction of the force experienced by a negative charge we use the "Right Hand Rule":

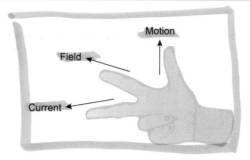

The Right Hand Rule for Forces on a negative charge in a magnetic field. A positive charge would experience a force in the opposite direction.

Top tip

A good way to remember this is that:

- The **First** finger represents the magnetic **Field**.

- The se**C**ond finger represents the electron **Current**.

- The **Th**umb represents **Th**rust or Motion (i.e. Force).

An even quicker way is to remember the American secret service **"FBI"** for the thumb (**F**), first finger (**B**) and second finger (**I**). Remember it is always the right hand used for electron current.

The relationship given in Equation 3.1 shows the two conditions that must be met for a charge to experience a force in a magnetic field. These are:

1. The charge must be moving.

2. The velocity of the moving charge must have a component perpendicular to the direction of the magnetic field.

Figure 3.2:

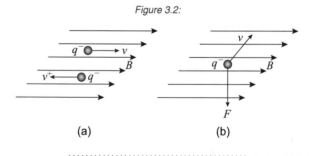

(a)   (b)

If the charge is moving parallel or antiparallel to the field, as shown in Figure 3.2(a), then no magnetic force acts on the moving charge and its velocity is unchanged. If the charge is moving with a velocity that is perpendicular to the direction of the magnetic field, as shown in Figure 3.2(b), then the magnetic force experienced by the charge is a maximum. In this last case, $\sin\Theta = \sin 90° = 1$ and so Equation 3.1 becomes

$$F = qvB$$

(3.2)

. . . .

Example

An electron enters a uniform magnetic field of magnetic induction 1.2 T with a velocity of 3.0×10^6 m s^{-1} at an angle of 30° to the direction of the magnetic field.

a) Calculate the magnitude of the magnetic force that acts on the electron.

b) If a proton instead of an electron enters the same magnetic field with the same velocity, what difference, if any, would there be to the force experienced by the proton?

a) Using Equation 3.1 we have

$$F = Bqv\sin\theta$$
$$\therefore F = 1.2 \times 1.6 \times 10^{-19} \times 3.0 \times 10^6 \times \sin 30°$$
$$\therefore F = 2.9 \times 10^{-13} \text{ N}$$

b) Since the charge on the proton has the same magnitude as the charge on the electron, the force on the proton also has the same magnitude.
The sense of the direction of the force is different, since a proton has a positive charge while an electron has a negative charge.

. . . .

3.3 The path of a charged particle in a magnetic field

We have seen that if a charged particle enters a magnetic field in a direction perpendicular to the magnetic field, then the field exerts a force on the particle in a direction that is perpendicular to both the field and the velocity of the particle. Because the magnetic force is always perpendicular to the velocity, v, of the particle, this force cannot change the *magnitude* of the velocity, only its *direction*. Since the magnetic force never has a component in the direction of v, then it follows that this force does no work on the moving charge. In other words, the *speed* of a charged particle moving in a magnetic field does not change although its *velocity* does because of the change in direction.

Figure 3.3: Charged particle in a magnetic field

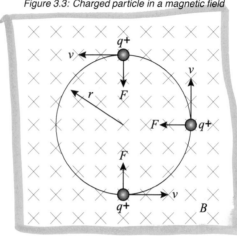

Consider Figure 3.3 showing a uniform magnetic field directed into the viewing plane. (Think of the field as looking at the end of an arrow travelling away from you.) A positively charged particle that, at the bottom of the figure, has a velocity from left to right will experience a force directed to the top of the figure, as shown. Although this force has no effect on the speed of the particle, it does change its direction. Since the constant-magnitude magnetic force always acts at right angles to the velocity, it can be seen that the particle follows a circular path, with the force always directed towards the centre of the circle.

We have met this type of motion before, where the force is always perpendicular to the direction of motion. It is similar to the force exerted by a rope on an object that is swung in a circle above the head. It is also the same as the gravitational force that keeps a planet or a satellite in orbit around its parent body.

The centripetal or radial acceleration of the particle is given by v^2/r where r is the radius of the circular path of the particle.

Using Newton's second law, we have

$$F = ma$$
$$\therefore qvB = m \times \frac{v^2}{r}$$
$$\therefore qB = \frac{mv}{r}$$

(3.3)

. . . .

where m is the mass of the particle.

We can rearrange Equation 3.3 to find the radius of the circular path.

$$qB = \frac{mv}{r}$$
$$\therefore r = \frac{mv}{qB}$$

(3.4)

. . . .

Example

An electron enters a uniform magnetic field of magnetic induction 1.2 T with a velocity of 3.0 x 10⁶ m s⁻¹ perpendicular to the direction of the magnetic field.

a) Calculate the radius of the circular path followed by the electron.

b) If a proton instead of an electron enters the same magnetic field with the same velocity, what difference, if any, would there be to the path followed by the proton?

a) Using Equation 3.4, we have for the electron

$$r_e = \frac{m_e v_e}{q_e B}$$
$$\therefore r_e = \frac{9.11 \times 10^{-31} \times 3 \times 10^6}{1.6 \times 10^{-19} \times 1.2}$$
$$\therefore r_e = 1.4 \times 10^{-5} \text{ m}$$

b) For the proton

$$r_p = \frac{m_p v_p}{q_p B}$$
$$\therefore r_p = \frac{1.67 \times 10^{-27} \times 3 \times 10^6}{1.6 \times 10^{-19} \times 1.2}$$
$$\therefore r_p = 0.026 \text{ m}$$

It can be seen that the radius of the circular path of the proton is much greater than that of the electron. This is because of the difference in masses of the two particles.

What is not shown by the calculation is that the particles will rotate in opposite directions because of the difference in the sign of their charges.

It is worth pointing out that Equation 3.4 only holds for charges moving with non-relativistic velocities. A value of v of about 10% of c is usually taken as the limit for the validity of this expression.

....

Charges entering a magnetic field

Go online

At this stage there is an online activity which shows the paths taken by charges that enter a uniform magnetic field perpendicular to the direction of the field.

This activity shows the paths taken by charges that enter a uniform magnetic field perpendicular to the direction of the field.

....

Quiz: The force on a moving charge

Go online

Useful data:

Fundamental charge e	1.6×10^{-19} C
Mass of an electron m_e	9.11×10^{-31} kg
Mass of proton	1.67×10^{-27} kg

Q1: A proton and an electron both enter the same magnetic field with the same velocity.

Which statement is correct?

a) Both particles experience the same magnetic force.
b) The particles experience the same magnitude of force but in opposite directions.
c) The proton experiences a larger force than the electron, in the same direction.
d) The proton experiences a smaller force than the electron, in the same direction.
e) The proton experiences a larger force than the electron, in the opposite direction.

....

Q2: An alpha particle that has a charge of $+2e$ enters a uniform magnetic field of magnitude 1.5 T, with a velocity of 5×10^5 m s^{-1}, perpendicular to the field.

What is the magnitude of the force on the particle?

a) 2.5×10^{-21} N
b) 1.3×10^{-15} N
c) 2.4×10^{-13} N
d) 1.2×10^{-13} N
e) 1.2×10^{-7} N

. . . .

Q3: Which of these bodies can move through a magnetic field without experiencing any net magnetic force?

i   a negatively-charged electron

ii   a positively-charged proton

iii an uncharged billiard ball

a) (i) only
b) (ii) only
c) (iii) only
d) none of them
e) all of them

. . . .

Q4: A charged particle moves in a magnetic field only.

Which statement is **always** true of the motion of this particle?

The particle moves with

a) constant speed.
b) constant velocity.
c) zero acceleration.
d) increasing acceleration.
e) decreasing acceleration.

. . . .

Q5: A beam of electrons is bent into a circle of radius 3.00 cm by a magnetic field of magnitude 0.60 mT.

What is the velocity of the electrons?

a) 3.16×10^6 m s^{-1}
b) 1.98×10^5 m s^{-1}
c) 1.72×10^3 m s^{-1}
d) 1.08×10^2 m s^{-1}
e) 2.88×10^{-24} m s^{-1}

. . . .

3.4   Helical motion in a magnetic field

Consider again the situation we looked at in previous Topic, where a charged particle q moves with a velocity v at an angle θ to a magnetic field B. We showed that the particle experiences a force F of magnitude $F = B\,q\,v\sin\theta$. This force is perpendicular to the plane containing B and v and so causes circular motion.

In that analysis, we ignored the component of v that is parallel to the magnetic field B. This component is $v \cos \theta$ as shown in Figure 3.4. In this figure, B and v are in the x-z plane, with B in the x-direction in this plane.

Figure 3.4: Charged particle moving in a magnetic field

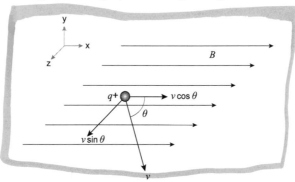

The component of the particle's velocity in the z-direction, $v \sin \theta$, perpendicular to the magnetic field that causes the circular motion, keeps the same magnitude but constantly changes direction. The component of the velocity in the x-direction, $v \cos \theta$, parallel to the magnetic field, remains constant in magnitude and direction because there is no force parallel to the field. The resulting motion of the charged particle is made up of two parts - circular motion with constant angular velocity at right angles to the magnetic field, and linear motion with constant linear velocity parallel to the magnetic field. This causes the particle to follow a helical path, with the axis of the **helix** along the direction of the magnetic field. This is shown in Figure 3.5.

Figure 3.5: Helical path of a charged particle in a magnetic field

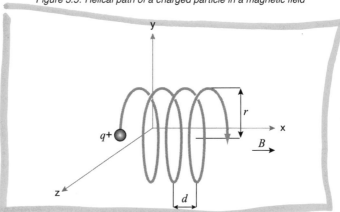

. . . .

The radius of the circular path of the particle is given by

$$r = \frac{mv \sin \theta}{qB}$$

(3.5)

. . . .

This is the equation derived in topic 6 using the component of the velocity perpendicular to the field, $v \sin \theta$.

The **pitch**, d, of the helix, or the distance travelled in the direction of the magnetic field per revolution, can be found as follows. (Remember from Unit 1, that periodic time T is the time taken to complete one revolution and it can be found from $\omega = \frac{2\pi}{T}$)

The speed at which the particle is moving in the x-direction is $v \cos \theta$, so

$$d = v \cos \theta \times T$$
$$\therefore d = v \cos \theta \times \frac{2\pi}{\omega}$$
$$\therefore d = v \cos \theta \times \frac{2\pi r}{v \sin \theta}$$
$$\therefore d = \frac{2\pi r}{\tan \theta}$$

(3.6)

. . . .

Example

A proton moving with a constant velocity of 3.0 x 10^6 m s^{-1} enters a uniform magnetic field of magnitude 1.5 T. The path of the proton makes an angle of 30° with the magnetic field as shown in Figure 3.6.

Figure 3.6: Path of a proton

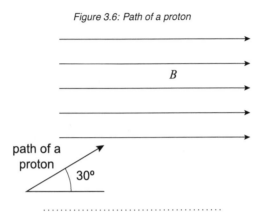

a) Describe and explain the shape of the path of the proton in the magnetic field.

b) Calculate the radius of the path of the proton.

c) Calculate the distance travelled by the proton in the direction of the magnetic field during one revolution.

a) The path is a helical shape.

This is because the proton has both circular and linear motion in the magnetic field. The circular motion is due to the component of its velocity $v \sin 30°$ perpendicular to the field. The linear motion exists because of the component v $\cos 30°$ parallel to the field.

b) To find the radius

$$r = \frac{mv \sin \theta}{qB}$$
$$\therefore r = \frac{1.67 \times 10^{-27} \times 3.0 \times 10^6 \times \sin 30}{1.6 \times 10^{-19} \times 1.5}$$
$$\therefore r = 0.010 \text{ m}$$

c)

To find the distance travelled per revolution

$$\text{distance travelled (pitch)} = \frac{2\pi r}{\tan \theta}$$
$$= \frac{2 \times \pi \times 0.010}{\tan 30}$$
$$= 0.11 \text{ m}$$

The path of a charged particle

At this stage there is an online activity which provides extra practice.

Go online

....

3.4.1 Charged particles in the Earth's magnetic field

We have already seen that there is a non-uniform magnetic field around the Earth, caused by its molten metallic core. This magnetic field is stronger near to the poles than at the equator. The Sun emits charged particles, both protons and electrons. Some of these charged particles enter the Earth's magnetic field near to the poles and in doing so, they spiral inwards. The magnetic field of the Earth traps these charged particles in regions known as the **Van Allen radiation belts** (Figure 3.7). Protons with relatively high masses are trapped in the inner radiation belts, while the electrons, with lower mass and greater speed, are trapped in the outer radiation belts.

Figure 3.7: The Van Allen radiation belts

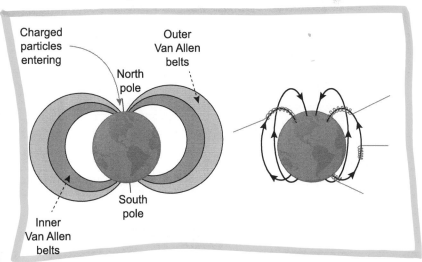

....

When they enter the Earth's atmosphere the charged particles can collide with the gases in the atmosphere. In doing so, some of their energy is emitted in the form of light. The light emitted from excited oxygen atoms is green in colour and that emitted from excited nitrogen atoms is pinkish-red. The resulting dramatic displays of coloured lights often seen dancing in the northern and southern night skies are called aurora - aurora borealis or the 'Northern Lights' and aurora australis or the 'Southern Lights'.

3.5 The Solar Wind

The solar wind is made up from a stream of plasma released from the upper layers of the Sun's atmosphere. Plasma is considered to be a fourth state of matter, where the electrons have detached from the atoms leaving a cloud of protons, neutrons and electrons and allowing the plasma to move as a whole. This upper layer of the Sun's atmosphere is correctly called the Corona which can only be seen with the naked eye during a solar eclipse.

The solar wind consists of mostly electrons and protons with energies usually between 1.5 and 10 keV. (An eV is an electrical unit of energy which is covered in more detail in Unit 3). These charged particles can only escape the immense pull of the Sun's gravity because of their high energy caused by the high temperature of the corona and the sun's powerful magnetic field.

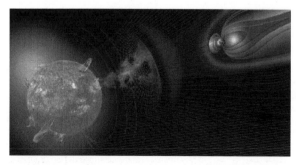

The solar wind flows from the sun and can travel great distances at speeds as high as 900 km s^{-1}. The region the solar wind occupies is known as the **heliosphere**.

The solar wind causes the tails of comets to trail behind the bodies of comets as they go through the solar system.

Comet Hale-Bopp. Photo Credit: A. Dimai and D. Ghirardo, (Col Druscie Obs.), AAC

A comet starts out as a lump of rock and ice contained in a region called an Oort Cloud but when the gravitational field of a nearby star such as the Sun attracts the comet, the heat of the star melts some of the ice contained in the comet which creates a gaseous tail that extends away from the source of the heat due to the solar wind. As a comet gets closer to the Sun it increases in speed due to the ever increasing gravitational field and as more ice evaporates the comet's tail will grow in length.

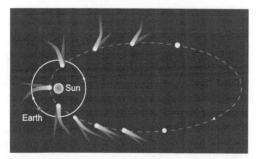

The upper atmosphere of the sun is incredibly unstable and often intense variations in brightness are seen, called a **solar flare**. A solar flare occurs when the magnetic energy that has built up in the solar atmosphere is suddenly released throwing radioactive material into space and releasing radiation across the full range of frequencies contained in the electromagnetic spectrum. The Sun's surface has huge magnetic loops called **prominences**.

These can cause geomagnetic storms which are induced by coronal mass ejections (CMEs) which have a big impact on equipment and astronauts in space causing effects such as corrosion of equipment, overloading of observation cameras and also exposing astronauts to dangerous levels of radiation. However, at the surface of the Earth we are well protected from the effects of solar flares and other solar activity by the Earth's magnetic field and atmosphere. See Figure 3.8 below.

Figure 3.8: Variation of cosmic rays with altitude

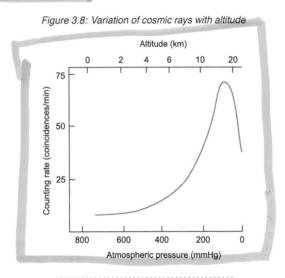

Like all planets with magnetic fields the Earth's magnetic field will interact with the solar wind to deflect the charged particles and form an elongated cavity in the solar wind. This cavity is called the **magnetosphere** of the planet. Near to the Earth the particles of the solar wind are travelling about 400 km s^{-1} as they are slowed by the magnetic field to produce a bow shaped shock wave around the earth.

Quiz: Motion of charged particles in a magnetic field

Go online

Useful data:

Fundamental charge e	1.6×10^{-19} C
Mass of an electron m_e	9.11×10^{-31} kg

Q6: A charged particle moving with a velocity v in a magnetic field B experiences a force F.

Which of the quantities v, B and F are *always* at right angles to each other?
(i) F and B
(ii) F and v
(iii) B and v

a) (i) only
b) (ii) only
c) (iii) only
d) (i) and (ii) only
e) (i) and (iii) only

....

Q7: A charged particle enters a magnetic field at an angle of 60° to the field.

Which is the best description of the resulting path of the charged particle through the field?

a) circular, parallel to the magnetic field
b) circular, perpendicular to the magnetic field
c) helical, with the axis in the magnetic field direction
d) helical, with the axis at 60° to the magnetic field
e) straight through the field

....

Q8: An electron moving with a constant velocity of 2.0×10^6 m s^{-1} enters a uniform magnetic field of value 1.0 T. The path of the electron makes an angle of 30° with the direction of the magnetic field.

What is the radius of the resulting path of the electron?

a) 8.24×10^{-5} m
b) 1.97×10^{-5} m
c) 1.14×10^{-5} m
d) 6.56×10^{-6} m
e) 5.69×10^{-6} m

....

Q9: What is the name given to the regions around the Earth where charged particles can become trapped?

a) aurora
b) ionosphere
c) northern lights
d) stratosphere
e) Van Allen belts

....

3.6 Extended information

Web links

There are web links available online exploring the subject further.

....

3.7 Summary

Summary

You should now be able to:

- explain what is meant by magnetic field and magnetic force;

- explain the behaviour of charged particles in a magnetic field;

- have an understanding of cosmic rays and the solar wind and how they interact with the Earth's atmosphere and magnetic field.

3.8 Assessment

End of topic 3 test

The following test contains questions covering the work from this topic.

Go online

  The following data should be used when required:

Fundamental charge e	1.6×10^{-19} C
Mass of an electron m_e	9.11×10^{-31} kg

Q10: A proton enters a magnetic field of magnitude 1.85×10^{-3} T with a velocity of 1.32×10^6 m s^{-1} at an angle of 49.5°.

Calculate the magnitude of the force on the proton.

_____ N

....

Q11: A beam of electrons travelling at 1.28×107 m s^{-1} enters a magnetic field of magnitude 3.62 mT, perpendicular to the field.

1. Calculate the force on an electron while it is moving in the magnetic field.
   _____ N
2. Calculate the radius of the circular path of the electrons in the field.
   _____ mm

....

Q12: A charged particle has a charge-to-mass ratio of 2.22×10^8 C kg^{-1}.

It moves in a circular path in a magnetic field of magnitude 0.713 T.

Calculate how long it takes the particle to complete one revolution.

_____ seconds

....

Q13: An ion of mass 6.6×10^{-27} kg, moving at a speed of 1×10^7 m s^{-1}, enters a uniform magnetic field of induction 0.67 T at right angles to the field.

The ion moves in a circle of radius 0.31 m within the magnetic field.

What is the charge on the ion, in terms of the charge on an electron?

_____

....

Q14: An electron travelling at velocity 4.9×10^5 m s^{-1} enters a uniform magnetic field at an angle of $50°$, as shown below.

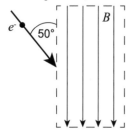

The electron travels in a helical path in the magnetic field, which has induction B = 4.5 $\times 10^{-3}$ T.

1. Calculate the radius of the helix.

   _ _ _ _ _ _ _ _ _ _ _ m
2. Calculate the pitch of the helix.

   _ _ _ _ _ _ _ _ _ _ _ m

. . . .

Q15: Which of the following statements are correct?

A) As a comet approaches a star it slows down due the gravitational field of the star.

B) As a comet approaches a star the comet's tail points away from the star due the solar wind.

C) As a comet approaches a star the comet's tail points towards the star due to the solar wind.

D) As a comet approaches a star the comet enters the heliosphere of the star.

E) As a comet approaches a star the comet's tail increases in size.

. . . .

Q16: What is the name given to the fourth state of matter?

a) bubbles
b) gas
c) liquid
d) plasma
e) solid

. . . .

Topic 4

Simple harmonic motion

Contents

Prerequisite knowledge

- *Newton's laws of motion.*

- *Radian measurement of angles and angular velocity (Unit 1 - Topic 2).*

- *Differentiation of trig functions.*

Learning objectives

By the end of this topic you should be able to:

- *understand what is meant by simple harmonic motion (SHM);*

- *apply the equations of motion in different SHM systems;*

- *understand how energy is conserved in an oscillating system;*

- *explain what is meant by damping of an oscillating system.*

4.1   Introduction

In most of the Unit 1 topics, we were studying motion where the acceleration was constant, whether it be linear motion or circular motion. In this topic we are looking at a specific situation in which the acceleration is not constant. We will be studying simple harmonic motion (SHM), in which the acceleration of an object depends on its displacement from a fixed point.

The topic begins by defining what is meant by simple harmonic motion, followed by derivations of the equations of motion for SHM. We then consider the kinetic energy and potential energy of an SHM system, as well as examples of different SHM systems. We will return to the subject of energy in the final section of the Topic, when we consider what happens to a 'damped' system.

4.2   Defining SHM

We will begin this topic by looking at the horizontal motion of a mass attached to a spring.

Figure 4.1: Mass attached to a horizontal spring

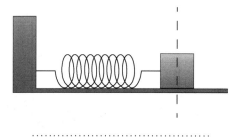

The mass rests on a frictionless surface, and the spring is assumed to have negligible mass. The spring obeys Hooke's law, so the force F required to produce an extension y of the spring is

$$F = ky$$

(4.1)

k is the spring constant, measured in N m^{-1}. This expression generally holds true for a spring so long as the extension is not large enough to cause a permanent deformation of the spring. If we pull the mass back a distance y and hold it (Figure 4.2), Newton's first law tells us the tension T in the spring (the restoring force) must also have magnitude F. In fact, $T = -F$, since force is a vector quantity, and T and F are in opposite directions. Hence,

$$T = -ky$$

(4.2)

. . . .

Figure 4.2: Mass pulled back and held

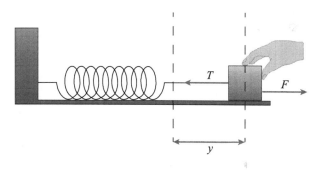

If we now release the mass (Figure 4.3), it is no longer in equilibrium, and we can apply Newton's second law $F = ma$. In this case the force acting on the block is T, so

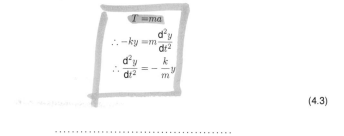

$$T = ma$$
$$\therefore -ky = m\frac{\mathrm{d}^2y}{\mathrm{d}t^2}$$
$$\therefore \frac{\mathrm{d}^2y}{\mathrm{d}t^2} = -\frac{k}{m}y$$

(4.3)

. . . .

Figure 4.3: Mass accelerated by the tension in the extended spring

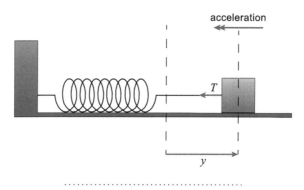

Equation 4.3 tells us that the acceleration of the mass is proportional to its displacement, since m and k are constants. The minus sign means that the acceleration and displacement vectors are in the opposite direction. This equation holds true whatever the extension or compression of the spring, since the spring also obeys Hooke's law when it is compressed (Figure 4.4).

Figure 4.4: Mass accelerated by the force in the compressed spring

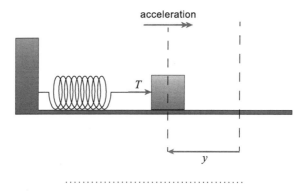

The mass will oscillate around the equilibrium ($T = 0$) position, with an acceleration always proportional to its displacement, and always in the opposite direction to its displacement vector. Such motion is called **simple harmonic motion (SHM)**.

A plot of displacement against time for the mass on a spring shows symmetric oscillations about $y = 0$. The maximum value of the displacement of an object performing SHM is called the **amplitude** of the oscillations.

Figure 4.5: Displacement against time for an object undergoing SHM

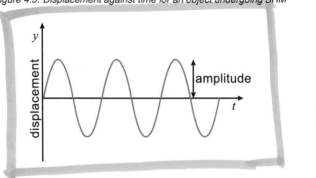

Equation 4.3 describes a specific example of SHM, that of a mass m attached to spring of spring constant k. In general, for an object performing SHM

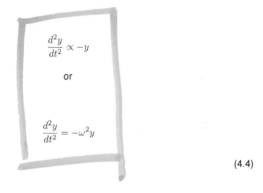

$$\frac{d^2y}{dt^2} \propto -y$$

or

$$\frac{d^2y}{dt^2} = -\omega^2 y$$

(4.4)

The constant of proportionality is ω^2, where ω is called the angular frequency of the motion. We are used to using ω to represent angular velocity in circular motion, and it is no coincidence that the same symbol is being used again. We will see now how simple harmonic motion and circular motion are related.

Example

A 1.0 kg mass attached to a horizontal spring (k = 5.0 N m^{-1}) is performing SHM on a horizontal, frictionless surface. What is the acceleration of the mass when its displacement is 4.0 cm?

Using Equation 4.3, the acceleration is:

$$a = -\frac{k}{m}y$$
$$\therefore a = -\frac{5.0}{1.0} \times 0.040$$
$$\therefore a = -0.20 \text{ m s}^{-2}$$

The magnitude of the acceleration is 0.20 m s⁻². Normally we do not need to include the minus sign unless we are concerned with the direction of the acceleration.

. . . .

4.3   Equations of motion in SHM

To try to help us understand SHM, we can compare the motion of an oscillating object with that of an object moving in a circle. Let us consider an object moving anti-clockwise in a circle of radius a at a constant angular velocity ω with the origin of an x-y axis at the centre of the circle. We will be concentrating on the y-component of the motion of this object. Assuming the object is at position $y = 0$ at time $t = 0$, then after time t the angular displacement of the object will be $\theta = \omega t$.

Figure 4.6: Object moving in a circle

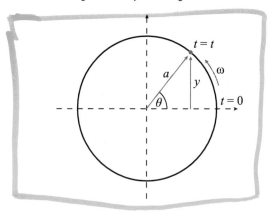

. . . .

From Figure 4.6, we can see that the y displacement at time t is equal to:

$$a \sin \theta = a \sin \omega t$$

Hence

$$y = a \sin \omega t$$

Differentiating this equation, the velocity in the *y*-direction is:

$$v = \frac{dy}{dt} = \omega a \cos \omega t$$

We can differentiate again to get the acceleration in the *y*-direction.

$$\frac{d^2y}{dt^2} = \frac{dv}{dt} = -\omega^2 a \sin \omega t$$
$$\therefore \frac{d^2y}{dt^2} = -\omega^2 y$$

(4.5)

Equation 4.5 gives us the relationship between acceleration and displacement, and is exactly the same as Equation 4.4 for an object undergoing SHM. (Remember it is the projection of the object on the *y*-axis that is undergoing SHM, rather than the object itself.) The angular velocity of the circular motion is therefore equivalent to the angular frequency of the SHM. As with circular motion, we can talk about the period of the SHM as the time taken to complete one oscillation. The relationship between the periodic time T and ω is:

$$T = \frac{2\pi}{\omega}$$

(4.6)

The **frequency** f of the SHM is the number of complete oscillations performed per second, so

$$f = \frac{1}{T} = \frac{\omega}{2\pi}$$
$$\therefore \omega = 2\pi f$$

(4.7)

Frequency is measured in hertz Hz, equivalent to s^{-1}. Another solution to Equation 4.4 is:

$$y = a \cos \omega t$$

The proof that $y = a \cos \omega t$ represents SHM is given in the following example.

Example

Prove that the equation $y = a \cos \omega t$ describes simple harmonic motion.

For an object to be performing SHM, it must obey the equation $d^2 y / dt^2 = -\omega^2 y$. We can differentiate our original equation twice:

$$y = a \cos \omega t$$
$$\therefore \frac{dy}{dt} = -\omega a \sin \omega t$$
$$\therefore \frac{d^2 y}{dt^2} = -\omega^2 a \cos \omega t$$
$$\therefore \frac{d^2 y}{dt^2} = -\omega^2 y$$

So $y = a \cos \omega t$ does represent SHM. In this case, the displacement is at a maximum when $t = 0$, whereas the displacement is 0 at $t = 0$ when $y = a \sin \omega t$.

There is one more equation we need to derive. We have expressed acceleration as a function of displacement (Equation 4.4 and Equation 4.5). We can also find an expression showing how the velocity varies with displacement. In deriving Equation 4.5 we used the expressions $y = a \sin \omega t$ and $v = \omega a \cos \omega t$. Rearranging both of these equations gives us

$$\sin \omega t = \frac{y}{a} \text{ and } \cos \omega t = \frac{v}{\omega a}$$

Now, using the trig identity $\sin^2\theta + \cos^2\theta = 1$, we have:

$$\sin^2\theta + \cos^2\theta = 1$$
$$\therefore \sin^2(\omega t) + \cos^2(\omega t) = 1$$
$$\therefore \frac{y^2}{a^2} + \frac{v^2}{\omega^2 a^2} = 1$$
$$\therefore y^2\omega^2 + v^2 = \omega^2 a^2$$
$$\therefore v^2 = \omega^2 a^2 - y^2\omega^2$$
$$\therefore v^2 = \omega^2 \left(a^2 - y^2\right)$$
$$\therefore v = \omega\sqrt{\left(a^2 - y^2\right)}$$

This equation tells us that for any displacement other than $y = a$, there are two possible values for v, since the square root can give positive or negative values, meaning that the velocity could be in two opposite directions. In fact, to emphasise this point the equation is usually written as,

$$v = \pm\omega\sqrt{a^2 - y^2}$$

(4.8)

....

Example

An object is performing SHM with amplitude 5.0 cm and frequency 2.0 Hz. What is the maximum value of the velocity of the object?

From Equation 4.8 and the on-screen animation, you should have realised that the velocity has its maximum value when the displacement is zero. So, remembering that $\omega = 2\pi f$.

$$v = \pm\,\omega\sqrt{a^2 - y^2}$$
$$\therefore v_{max} = 2\pi f \times \sqrt{a^2 - 0}$$
$$\therefore v_{max} = 2\pi f \times a$$
$$\therefore v_{max} = 2\pi \times 2.0 \times 0.050$$
$$\therefore v_{max} = 0.63 \text{ m s}^{-1}$$

....

Displacement, velocity and acceleration in SHM

At this stage there is an online activity which allows you to make a thorough investigation of the mass-on-a-spring system..

Go online

....

Quiz: Defining SHM and equations of motion

Go online

Q1: For an object to be performing simple harmonic motion, the net force acting on it must be proportional to its

a) frequency.
b) displacement.
c) velocity.
d) mass.
e) amplitude.

. . . .

Q2: An object is performing SHM with ω = 8.00 rad s^{-1} and amplitude a = 0.150 m. What is the frequency of the oscillations?

a) 1.27 Hz
b) 2.55 Hz
c) 25.1 Hz
d) 50.3 Hz
e) 120 Hz

. . . .

Q3: An object is performing SHM oscillations. Which one of these properties is continuously changing as the object oscillates?

a) Period
b) Mass
c) Amplitude
d) Acceleration
e) Frequency

. . . .

Q4: A mass on a spring oscillates with frequency 5.00 Hz and amplitude 0.100 m. What is the maximum acceleration of the mass?

a) 0.063 ms^{-2}
b) 2.50 ms^{-2}
c) 10.0 ms^{-2}
d) 24.7 ms^{-2}
e) 98.7 ms^{-2}

. . . .

Q5: An object is performing SHM with ω = 4.2 rad s^{-1} and amplitude 0.25 m. What is the speed of the object when its displacement is 0.15 m from its equilibrium position?

a) 0.17 ms^{-1}
b) 0.42 ms^{-1}
c) 0.84 ms^{-1}
d) 1.1 ms^{-1}
e) 3.5 ms^{-1}

. . . .

4.4   Energy in SHM

Let us return to the first example of SHM we looked at, a mass attached to a spring, oscillating horizontally on a smooth surface. We will now consider the energy in this system. The principle of conservation of energy means that the total energy of the system must remain constant.

As the displacement y of the mass increases, the velocity decreases, and so the kinetic energy of the mass decreases. To keep the total energy of the system constant, this 'lost' energy is stored as potential energy in the spring. The more the spring is stretched or compressed, the greater the potential energy stored in it. We can derive equations to show how the kinetic and potential energies of the system vary with the displacement of the mass.

We already know that the velocity of an object performing SHM is $v = \pm\omega\sqrt{a^2 - y^2}$. We also know that the kinetic energy of a object of mass m is $KE = \frac{1}{2}mv^2$. So

$$v^2 = \omega^2\left(a^2 - y^2\right)$$
$$\therefore KE = \frac{1}{2}m\omega^2\left(a^2 - y^2\right)$$

$$(4.9)$$

To obtain the potential energy, consider the work done against the spring's restoring force in moving the mass to a displacement y. We cannot use a simple *work done = force × distance* calculation since the force changes with displacement. Instead we must use a calculus approach, calculating the work done in increasing the displacement by a small amount $\mathrm{d}y$ and then performing an integration over the limits of y.

The work done in moving a mass to a displacement y is equal to:

$$\int_{y=0}^{y} F\mathrm{d}y$$

This work done is stored as potential energy in the system, so

$$PE = \int_{y=0}^{y} F\mathrm{d}y$$

Now, the general equation for SHM is $\mathrm{d}^2y/\mathrm{d}t^2 = -\omega^2 y$, and Newton's second law of motion tells us that *force = mass × acceleration*, so

$$F = m\frac{\mathrm{d}^2y}{\mathrm{d}t^2} = -m\omega^2 y$$

Substituting for F in the potential energy equation

$$PE = \int_{y=0}^{y} F \mathrm{d}y$$
$$\therefore PE = \int_{y=0}^{y} m\omega^2 y \mathrm{d}y$$

Note that the minus sign has been dropped. This is because work is done **on** the spring in expanding it from $y = 0$ to $y = y$. The potential energy stored in the spring is positive, since the PE is at its minimum value (zero) when $y = 0$

$$PE = \int_{y=0}^{y} m\omega^2 y \mathrm{d}y$$
$$\therefore PE = m\omega^2 \int_{y=0}^{y} y \mathrm{d}y$$
$$\therefore PE = \frac{1}{2}m\omega^2 y^2$$

(4.10)

. . . .

Energy in simple harmonic motion

Consider an SHM system of a 0.25 kg mass oscillating with $\omega = \pi$ rad s^{-1} and amplitude $a = 1.2$ m. Sketch graphs showing:

1. PE, KE and total energy against displacement.

2. PE, KE and total energy against time.

In both cases sketch the energy over three cycles, assuming the displacement $y = 0$ at $t = 0$.

. . . .

In summary then, the kinetic energy is at a maximum when the speed of an oscillating object is at a maximum, which is when the displacement y equals zero. Note that at $y = 0$, the force acting on the object is also zero, and hence the potential energy is zero. At the other extreme of the motion, when $y = a$, the velocity is momentarily zero, and so the kinetic energy is zero. The restoring force acting on the object is at a maximum at this point, and so the potential energy of the system is at a maximum. At all other points in the motion of the object, the total energy of the system is partly kinetic and partly potential, with the sum of the two being constant.

Energy of a mass on a spring

At this stage there is an online activity which allows you to see the energy-time graph plotted out as the mass oscillates.

Go online

. . . .

Quiz: Energy in SHM

Q6: A 2.0 kg block is performing SHM with angular frequency 1.6 rad s^{-1}. What is the potential energy of the system when the displacement of the block is 0.12 m?

Go online

a) 0.023 J
b) 0.037 J
c) 0.074 J
d) 0.19 J
e) 0.38 J

. . . .

Q7: When the kinetic energy of an SHM system is 25 J, the potential energy of the system is 60 J. What is the potential energy of the system when the kinetic energy is 40 J?

a) 15 J
b) 25 J
c) 37.5 J
d) 45 J
e) 60 J

. . . .

Q8: A mass attached to a spring is oscillating horizontally on a smooth surface. At what point in its motion does the kinetic energy have a maximum value?

a) When its displacement from the rest position is at its maximum.
b) When the mass is midway between its rest position and its maximum displacement.
c) When its displacement from the rest position is zero.
d) The kinetic energy is constant at all points in its motion.
e) Cannot tell without knowing the amplitude and frequency.

. . . .

Q9: A mass-spring system is oscillating with amplitude a. What is the displacement at the point where the kinetic energy of the system is equal to its potential energy?

a) $\pm a/4$
b) $\pm a/2$
c) $\pm a/\sqrt{2}$
d) $\pm a\sqrt{2}$
e) $\pm 2a$

. . . .

Q10: The total energy of a system oscillating with SHM is 100 J. What is the kinetic energy of the system at the point where the displacement is half the amplitude?

a) 8.66 J
b) 10 J
c) 50 J
d) 75 J
e) 100 J

....

4.5   Applications and examples

In this section we will cover following applications:

- Mass on a spring - vertical oscillations

- Simple pendulum

- Loudspeaker cones and eardrums.

4.5.1   Mass on a spring - vertical oscillations

We have been using a mass on a spring to illustrate different aspects of SHM. In Section 3.2 we analysed the horizontal mass-spring system. We will look now at the vertical mass-spring system. This is the system we used in the on-screen activities earlier in this Topic. Here we will analyse the forces acting on the mass to show that its motion is indeed SHM.

Figure 4.7: Mass hanging from a spring

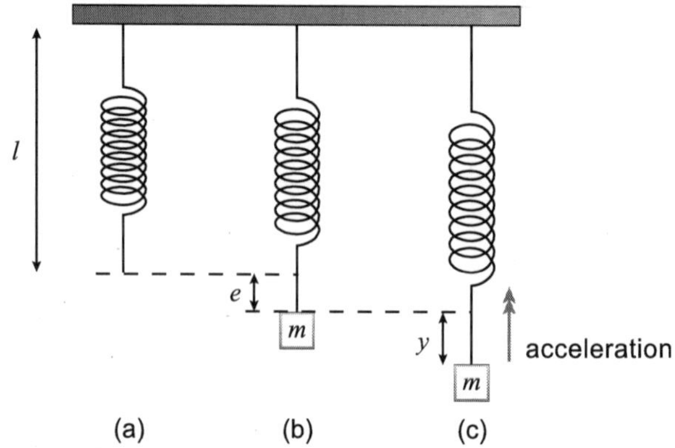

(a)   (b)   (c)

....

The spring has a natural length l, spring constant k and negligible mass Figure 4.7(a). When a mass m is attached to the spring Figure 4.7(b), it causes an extension e, and the system is at rest in equilibrium. Hence the tension T in the spring ($k \times e$) acting upwards on the mass must be equal in magnitude and opposite in direction to the weight mg. If the mass is now pulled down a distance y and released Figure 4.7(c), the tension in the spring is $k(y + e)$. The resultant force acting upwards is equal to $T - W$. Hence the resultant force is

$$F = k(y + e) - mg$$
$$\therefore F = ky + ke - mg$$

But $ke = mg$, so the resultant force $F = ky$. Using Newton's second law, this resultant force must be equal to (*mass* × *acceleration*), so

$$ky = -m\frac{d^2 y}{dt^2}$$
$$\therefore \frac{d^2 y}{dt^2} = -\frac{k}{m}y$$

(4.11)

. . . .

The minus sign appears because the displacement and acceleration vectors are in opposite directions. Equation 4.11 has the form ${d^2 y}/{dt^2} = -\omega^2 y$ showing that the motion is SHM, with

$$\omega = \sqrt{\frac{k}{m}}$$

4.5.2 Simple pendulum

Another SHM system is the simple pendulum. A simple pendulum consists of a bob, mass m, attached to a light string of length l. The bob is pulled to one side through a small angle θ and released. The subsequent motion is SHM.

Figure 4.8: Simple pendulum

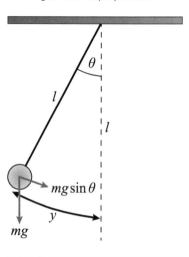

The restoring force on the bob acts perpendicular to the string, so no component of the tension in the string contributes to the restoring force. The restoring force is $mg \sin \theta$, the component of the weight acting perpendicular to the string. Now for small θ, when θ is measured in radians, $\sin \theta \approx \theta$. Also, since we are measuring θ in radians, $\theta = y/l$. Hence the restoring force acting on the bob is $mg \sin \theta = mg \times y/l$.

Now, using Newton's second law

$$mg\frac{y}{l} = -m\frac{\mathrm{d}^2 y}{\mathrm{d}t^2}$$
$$\therefore \frac{\mathrm{d}^2 y}{\mathrm{d}t^2} = -\frac{g}{l}y$$

(4.12)

Again, the minus sign appears because the displacement and acceleration vectors are in the opposite directions. So we have SHM with $\omega = \sqrt{g/l}$. Note that the angular frequency, and hence the period, are both **independent of the mass of the bob**. Note also that we have assumed small θ and hence small displacement y, in deriving Equation 4.12. The simple pendulum only approximates to SHM because of the approximation $\sin \theta \approx \theta$.

Example

Two simple pendula are set in motion. Pendulum A has string length p and a bob of mass q. Pendulum B has string length $4p$ and the mass of its bob is $5q$. What is the ratio T_A/T_B of the periodic times of the two pendula?

A simple pendulum has $\omega = \sqrt{g/l}$, so the periodic time

$$T = \frac{2\pi}{\omega} = 2\pi\sqrt{\frac{l}{g}}$$

The ratio of periodic times is therefore

$$\frac{T_A}{T_B} = \frac{2\pi\sqrt{\frac{l_A}{g}}}{2\pi\sqrt{\frac{l_A}{g}}}$$

$$\therefore \frac{T_A}{T_B} = \frac{\sqrt{l_A}}{\sqrt{l_B}}$$

$$\therefore \frac{T_A}{T_B} = \frac{\sqrt{p}}{\sqrt{4p}}$$

$$\therefore \frac{T_A}{T_B} = \frac{1}{2}$$

So the ratio of periodic times depends on the square root of the ratio of string lengths. Remember, the mass of the bob does not affect the periodic time of a pendulum.

. . . .

Simple pendulum

At this stage there is an online activity which explores the oscillations of a simple pendulum and demonstrates why this is SHM.

Go online

. . . .

Understanding, significance and treatment of uncertainties

At this stage there is an online activity which will help you understand how to analyse and interpret uncertainties.

Go online

. . . .

4.5.3 Loudspeaker cones and eardrums

A practical situation in which SHM occurs is in the generation and detection of sound waves. Sound waves are longitudinal waves, in which air molecules are made to oscillate back and forth due to high and low pressure regions being created. In a loudspeaker, a cardboard cone is rigidly attached to an electric coil which sits in a magnetic field. A fluctuating electric current in the coil causes the coil and the cone to vibrate back and forth. A pure note causes SHM vibrations of the cone (other notes cause more complicated vibrations). The oscillations of the cone cause the air molecules to oscillate, creating the high and low pressure regions that cause the wave to travel.

The opposite process occurs in the ear, where the oscillations of air molecules cause the tympanic membrane (commonly called the eardrum) to oscillate. These oscillations are converted to an electrical signal in the inner ear, and are transmitted to the brain via the auditory nerve.

Quiz: SHM Systems

Go online

Q11: An SHM oscillator consists of a 0.500 kg block suspended by a spring (k = 2.00 N m^{-1}), oscillating with amplitude 8.00 × 10^{-2} m. What is the period T of this system?

a) 0.318 s
b) 1.25 s
c) 2.00 s
d) 3.14 s
e) 12.6 s

....

Q12: What is the frequency f of a simple pendulum consisting of a 0.250 kg mass attached to a 0.300 m string?

a) 0.0278 Hz
b) 0.910 Hz
c) 0.997 Hz
d) 5.20 Hz
e) 35.9 Hz

....

Q13: A 0.19 kg mass is oscillating vertically, attached to a spring. The period of the oscillations is 0.45 s. What is the spring constant of the spring?

a) 2.7 N m^{-1}
b) 5.9 N m^{-1}
c) 17 N m^{-1}
d) 28 N m^{-1}
e) 37 N m^{-1}

....

Q14: A simple pendulum is oscillating with period 0.75 s. If the pendulum bob, mass m, is replaced with a bob of mass $2m$, what is the new period of the pendulum?

a) 0.375 s
b) 0.75 s
c) 1.06 s
d) 1.5 s
e) 3.0 s

. . . .

Q15: A simple pendulum has frequency 5 Hz. How would you increase its frequency to 10 Hz?

a) Decrease its length by a factor of 4.
b) Decrease its length by a factor of 2.
c) Increase its length by a factor of 2.
d) Increase its length by a factor of 4.
e) Increase its length by a factor of 10.

. . . .

4.6 Damping

In the examples we have looked at so far, we have ignored any external forces acting on the system, so that the oscillator continues to oscillate with the same amplitude and no energy is lost from the system. In reality, of course, this does not happen. For our horizontal mass-spring system, for example, friction between the mass and the horizontal surface would mean that some energy would be 'lost' from the system with every oscillation.

At the maximum displacement, all of the system's energy is potential energy, given by Equation 4.10

$$PE = \tfrac{1}{2}m\omega^2 y^2$$

If the energy of the system is decreasing, the amplitude must also be decreasing to satisfy the above equation. The **damping** of the system describes the rate at which energy is being lost, or the rate at which the amplitude is decreasing.

A system in which the damping effects are small is described as having light damping. A plot of displacement against time for such an oscillator is shown in Figure 4.9

Figure 4.9: Lightly-damped system

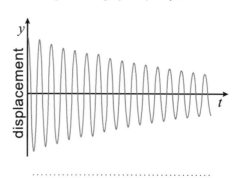

....

In a heavily damped system, the damping is so great that no complete oscillations are seen, and the 'oscillating' object does not travel past the equilibrium point.

Figure 4.10: Heavily-damped and critically-damped systems

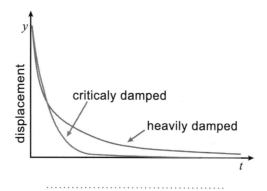

....

A critically damped system is one with an amount of damping that ensures the oscillator comes to rest in the minimum possible time. Another way of stating this is that critical damping is the minimum amount of damping that completely eliminates the oscillations.

An important application of damping is in the suspension system of a car. The shock absorber attached to the suspension spring consists of a piston in a reservoir of oil which moves when the car goes over a bump in the road. Underdamping (light damping) would mean that the car would continue to wobble up and down as it continued along the street. Overdamping (heavy damping) would make the suspension system ineffective, and produce a bumpy ride. Critical damping provides the smoothest journey, absorbing the bumps without making the car oscillate.

How shock absorbers work

. . . .

There is a link to the useful animation available online, where you can decide for yourself which type of damping is used in the shock absorber.

4.7 Extended information

Web links

There are web links available online exploring the subject further.

. . . .

4.8 Summary

A system is oscillating with simple harmonic motion if its acceleration is proportional to its displacement, and is always directed towards the centre of the motion. We have analysed several SHM systems, and looked at some applications of SHM. Conservation of energy has been discussed, and the effects of damping have been demonstrated.

Summary

You should now be able to:

- define what is meant by simple harmonic motion, and be able to describe some examples of SHM;

- state the equation $\frac{d^2 y}{dt^2} = -\omega^2 y$ and explain what each term in this equation means;

- show that $y = a \cos \omega t$ and $y = a \sin \omega t$ are solutions of the above equation, and show that $v = \pm \omega \sqrt{a^2 - y^2}$ in both of these cases;

- derive expressions for the potential and kinetic energies of an SHM system;

- explain what is meant by damping of an oscillating system.

4.9 Assessment

End of topic 4 test

The following test contains questions covering the work from this topic.

Go online

The following data should be used when required:

Acceleration due to gravity g	9.8 m s^{-2}

The end of topic test is available online. If however you do not have access to the web, you may try the following questions.

Q16: A 1.25 kg mass is attached to a light spring ($k = 40 \text{ N m}^{-1}$) and rests on a smooth horizontal surface. The mass is pulled back a distance of 3.2 cm and released.

Calculate the size of the acceleration of the mass at the instant it is released.

_____ m s^{-2}

....

Q17: A 0.46 kg mass attached to a spring ($k = 26 \text{ N m}^{-1}$) is performing SHM on a smooth horizontal surface.

Calculate the periodic time of these oscillations.

_____ s

....

Q18: An SHM system is oscillating with angular frequency $\omega = 3.6 \text{ rad s}^{-1}$ and amplitude $a = 0.24$ m.

1. Calculate the maximum value of the acceleration of the oscillator.
   _____ m s^{-2}
2. Calculate the maximum value of the velocity of the oscillator.
   _____ m s^{-1}

....

Q19: An SHM system consists of a 1.04 kg mass oscillating with amplitude 0.257 m.

If the angular frequency of the oscillations is 8.00 rad s^{-1}, calculate the total energy of the system.

_____ J

....

Q20: A body of mass 0.85 kg is performing SHM with amplitude 0.35 m and angular frequency 5.5 rad s^{-1}.

Calculate the displacement of the body when the potential energy of the system is equal to its kinetic energy.

_____ m

....

Q21: A simple pendulum has a bob of mass 0.55 kg and a string of length 0.47 m.

Calculate the frequency of the SHM oscillations of this pendulum.

_____ Hz

....

Q22: A simple pendulum, consisting of a 0.15 kg bob on a 0.92 m string, is oscillating with amplitude 44 mm.

Calculate the maximum kinetic energy of the pendulum.

_____ J

. . . .

Q23: An SHM system consists of an oscillating 1.5 kg mass. The system is set into motion, with initial amplitude 0.37 m. Damping of the system means that the mass continues to oscillate with an angular frequency of 4.0 rad s^{-1} but 10% of the system's energy is lost in work against friction with every oscillation.

Calculate the amplitude of the second oscillation after the system is set in motion.

_____ m

. . . .

Topic 5

Waves

Contents

Prerequisite knowledge

- *Radian measurement of angles.*
- *Simple harmonic motion.*

Learning objectives

By the end of this topic you should be able to:

- *use all the terms commonly employed to describe waves;*
- *derive an equation describing travelling sine waves, and solve problems using this equation;*
- *show an understanding of the difference between travelling and stationary waves;*
- *calculate the harmonics of a number of stationary wave systems;*
- *apply the principle of superposition;*
- *be able to explain how beats can be used to tune musical instruments;*
- *to understand the term phase difference and use the phase angle equation.*

5.1   Introduction

A wave is a travelling disturbance that carries energy from one place to another, but with no net displacement of the medium. You should already be familiar with transverse waves, such as light waves, where the oscillations are perpendicular to the direction in which the waves are travelling; and longitudinal waves, such as sound waves, where the oscillations of the medium are parallel to the direction in which the waves are travelling.

We begin this topic with a review of the basic definitions and terminology used to describe waves. We will discuss what terms such as 'frequency' and 'amplitude' mean in the context of light and sound waves. Section 3.3 deals with travelling waves, and we will derive and use mathematical expressions to describe these sorts of waves. Section 3.4 looks at the principle of superposition, which tells us what happens when two or more waves overlap at a point in space, and how they can be used in synthesisers. Section 3.5 looks at Stationary Waves and again we will derive and use mathematical expressions to describe these sorts of waves and look at their application in music. Section 3.6 looks at an interesting phenomenon in music called Beats which are used to tune musical instruments.

5.2   Definitions

The easiest way to explain wave phenomena is to visualise a train of waves travelling along a rope. The plot of displacement (y) against distance (x) is then exactly the same as the rope looks while the waves travel along it. We will be concentrating on sine and cosine waves as the most common sorts of waves, and the simplest to describe mathematically (see Figure 5.1).

Figure 5.1: Travelling sine wave

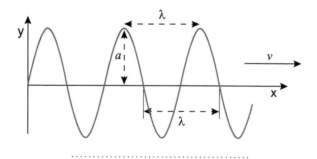

As a train of waves passes along the rope, each small portion of the rope is oscillating in the y-direction. There is no movement of each portion of the rope in the x-direction, and when we talk about the **speed** v of the wave, we are talking about the speed at which the disturbance travels in the x-direction.

The **wavelength** λ is the distance between two identical points in the wave cycle, such as the distance between two adjacent crests. The **frequency** f is the number of complete waves passing a point on the x-axis in a given time period. When this time period is one second, f is measured in hertz (Hz), equivalent to s^{-1}. 1 Hz is therefore equivalent to one complete wave per second. The relationship between these three quantities is

$$v = f\lambda$$

(5.1)

....

The **periodic time** T (or simply the **period** of the wave) is the time taken to complete one oscillation, in the same way that the periodic time we use to describe circular motion is the time taken to complete one revolution. The period is related to the frequency by the simple equation.

$$T = \frac{1}{f}$$

(5.2)

....

The **amplitude** a of the wave is the maximum displacement in the y-direction. As the waves pass along the rope, the motion of each portion of the rope follows the simple harmonic motion relationship

$$y = a\sin(2\pi ft)$$

We will use this expression to work out a mathematical relationship to describe wave motion later in this topic.

We normally use the wavelength to describe a light wave, or any member of the electromagnetic spectrum. The visible spectrum extends from around 700 nm (red) to around 400 nm (blue) (1 nm = 10^{-9} m). Longer wavelengths go through the infrared and microwaves to radio waves. Shorter wavelengths lead to ultraviolet, X-rays and gamma radiation (see Figure 5.2). The frequency of visible light is of the order of 10^{14} Hz (or 10^5 GHz, where 1 GHz = 10^9 Hz).

Figure 5.2: The electromagnetic spectrum

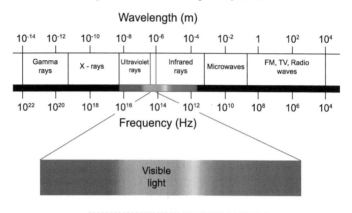

Sound waves are usually described by their frequency (or **pitch**). The human ear can detect sounds in the approximate range 20 Hz - 20 000 Hz. Sound waves with frequencies greater than 20 000 Hz are called ultrasonic waves, whilst those with frequencies lower than 20 Hz are infrasonic. The musical note middle C, according to standard concert pitch, has frequency 261 Hz.

The **irradiance** I of a wave tells us the amount of power falling on unit area, and is measured in W m^{-2}. The irradiance is proportional to a^2. In practical terms, this means the brightness of a light wave, or the loudness of a sound wave, depends on the amplitude of the waves, and increases with a^2.

Finally in this Section, let us consider the two waves in Figure 5.3.

Figure 5.3: Two out-of-phase sine waves

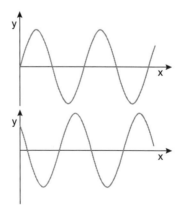

....

In terms of amplitude, frequency and wavelength, these waves are identical, yet they are 'out-of-step' with each other. We say they are **out of phase** with each other. A wave is an oscillation of a medium and the **phase** of the wave tells us how far through an oscillation a point in the medium is. Coherent waves **coherent waves** have the same speed and frequency (and similar amplitudes) and so there is a **constant phase relationship between two coherent waves.**

Quiz: Properties of waves

Go online

Useful data:	
Speed of light c	$3.00 \times 10^8 \ m \ s^{-1}$

Q1:   Which one of the following quantities should be increased to increase the volume (loudness) of a sound wave?

a) frequency
b) wavelength
c) speed
d) phase
e) amplitude

....

Q2:   The visible spectrum has the approximate wavelength range 400 - 700 nm. What is the approximate frequency range of the visible spectrum?

a) $2.1 \times 10^{14} - 1.2 \times 10^{15}$ Hz
b) $4.3 \times 10^{14} - 7.5 \times 10^{14}$ Hz
c) $1.3 \times 10^{15} - 2.3 \times 10^{15}$ Hz
d) $4.3 \times 10^{20} - 7.5 \times 10^{20}$ Hz
e) $1.3 \times 10^{20} - 2.3 \times 10^{20}$ Hz

....

Q3:   Which of these sets of electromagnetic waves are listed in order of **increasing** wavelength?

a) X-rays, infrared, microwaves.
b) radio waves, gamma rays, visible.
c) infrared, visible, ultraviolet.
d) X-rays, infrared, ultraviolet.
e) microwaves, visible, infrared.

....

Q4: What is the frequency of a beam of red light from a helium-neon laser, which has wavelength 633 nm?

a) 190 Hz
b) 2.11×10^5 GHz
c) 4.74×10^5 GHz
d) 2.11×10^8 GHz
e) 4.74×10^8 GHz

....

Q5: A laser produces light waves of average amplitude a m. The irradiance of the beam is 20 W m^{-2}. What is the irradiance if the average amplitude is increased to $3a$ m?

a) 6.7 W m^{-2}
b) 23 W m^{-2}
c) 60 W m^{-2}
d) 180 W m^{-2}
e) 8000 W m^{-2}

....

5.3   Travelling waves

In this section, we will attempt to find a mathematical expression for a **travelling wave**. The example we shall consider is that of a train of waves being sent along a rope in the x-direction, but the expression we will end up with applies to all transverse travelling waves.

We will start by considering what happens to a small portion of the rope as the waves travel through it. From the definition of a wave we know that although the waves are travelling in the x-direction, there is no net displacement of each portion of the rope in that direction. Instead each portion is performing simple harmonic motion (SHM) in the y-direction, about the $y = 0$ position, and the SHM of each portion is slightly out of step (or phase) with its neighbours.

The displacement of one portion of the rope is given by the SHM equation

$$y = a \sin(2\pi f t)$$

where y is the displacement of a particle at time t, a is the amplitude and f is the frequency of the waves.

The wave disturbance is travelling in the x-direction with speed v. Hence at a distance x from the origin, the disturbance will happen after a time delay of x/v. So the disturbance at a point x after time t is exactly the same as the disturbance at the point $x = 0$ at time $(t - x/v)$.

We can therefore find out exactly what the disturbance is at point x at time t by replacing t by $(t - x/v)$ in the SHM equation

$$y = a \sin 2\pi f \left(t - \frac{x}{v} \right)$$

We can re-arrange this equation, substituting for $v = f\lambda$

$$y = a \sin 2\pi f \left(t - \frac{x}{v} \right)$$
$$\therefore y = a \sin 2\pi \left(ft - \frac{fx}{v} \right)$$
$$\therefore y = a \sin 2\pi \left(ft - \frac{fx}{f\lambda} \right)$$
$$\therefore y = a \sin 2\pi \left(ft - \frac{x}{\lambda} \right)$$

(5.3)

. . . .

Note that we are taking the sine of the angle $2\pi \left(ft - x/\lambda \right)$. This angle is expressed in **radians**. You should also note that this expression assumes that at $t = 0$, the displacement at $x = 0$ is also 0.

For a wave travelling in the -x direction, Equation 5.3 becomes

$$y = a \sin 2\pi \left(ft + \frac{x}{\lambda} \right)$$

(5.4)

. . . .

We can now calculate the displacement of the rope, or any other medium, at position x and time t if we know the wavelength and frequency of the waves.

Travelling waves and the wave equation

Go online

At this stage there is an online activity which explores how the appearance and speed of a travelling wave depend on the different parameters in the wave equation.

This activity explores how the appearance and speed of a travelling wave depend on the different parameters in the wave equation.

. . . .

Example A periodic wave travelling in the x-direction is described by the equation

$$y = 0.2 \sin (4\pi t - 0.1x)$$

What are (a) the amplitude, (b) the frequency, (c) the wavelength, and (d) the speed of the wave? (All quantities are in S.I. units.)

To obtain these quantities, we first need to re-arrange the expression for the wave into a form more similar to the general expression given for a travelling wave. The general expression is

$$y = a \sin 2\pi \left(ft - \frac{x}{\lambda} \right)$$

We are given

$$y = 0.2 \sin (4\pi t - 0.1x)$$

Re-arranging

$$y = 0.2 \sin 2\pi \left(2t - \frac{0.1x}{2\pi} \right)$$

So by comparison, we can see that:

a) the amplitude a = 0.2 m

b) the frequency f = 2 Hz

c) the wavelength $\lambda = {}^{2\pi}/_{0.1} = 63$ m

d) By calculation, the speed of the wave $v = f\lambda = 2 \times 63 = 130$ m s^{-1}

. . . .

Example Consider again the travelling wave in the previous example, described by the equation

$$y = 0.2 \sin (4\pi t - 0.1x)$$

Calculate the displacement of the medium in the y-direction caused by the wave at the point $x = 25$ m when the time $t = 0.30$ s.

To calculate the y-displacement, put the data into the travelling wave equation. Remember to take the sine of the angle measured in *radians*.

$$y = 0.2 \sin (4\pi t - 0.1x)$$
$$\therefore y = 0.2 \sin ((4 \times \pi \times 0.30) - (0.1 \times 25))$$
$$\therefore y = 0.2 \sin 1.2699$$
$$\therefore y = 0.2 \times 0.9551$$
$$\therefore y = 0.19 \,\text{m}$$

....

Quiz: Travelling waves

Q6: The equation representing a wave travelling along a rope is

Go online

$$y = 0.5 \sin 2\pi \left(0.4t - \frac{x}{12}\right)$$

At time $t = 2.50$ s, what is the displacement of the rope at the point $x = 7.00$ m?

a) 0.00 m
b) 0.23 m
c) 0.25 m
d) 0.40 m
e) 0.50 m

....

Q7: Waves are being emitted in the x-direction at frequency 20 Hz, with a wavelength of 1.0 m. If the displacement at $x = 0$ is 0 when $t = 0$, which one of the following equations could describe the wave motion?

a) $y = \sin 2\pi (t - 20x)$
b) $y = 2 \sin 2\pi (20t - x)$
c) $y = \sin 2\pi \left(\frac{t}{20} - x\right)$
d) $y = 3 \cos 2\pi (t - 20x)$
e) $y = 20 \sin 2\pi (t - 20x)$

....

Q8: A travelling wave is represented by the equation

$$y = 4 \cos 2\pi (t - 2x)$$

What is the displacement at $x = 0$ when $t = 0$?

a) 0 m
b) 1 m
c) 2 m
d) 3 m
e) 4 m

. . . .

Q9:   A travelling wave is represented by the equation

$$y = \sin 2\pi \left(12t - 0.4x\right)$$

What is the frequency of this wave?

a) 0.4 Hz
b) 0.833 Hz
c) 2.0 Hz
d) 2.5 Hz
e) 12 Hz

. . . .

Q10: A travelling wave is represented by the equation

$$y = 2 \sin 2\pi \left(5t - 4x\right)$$

What is the speed of this wave?

a) 0.80 m s^{-1}
b) 1.25 m s^{-1}
c) 4.0 m s^{-1}
d) 5.0 m s^{-1}
e) 20 m s^{-1}

. . . .

5.4   Superposition of waves and phase difference

In this topic we will cover the following:

• The principle of superposition
• The fourier series

5.4.1 Principle of superposition

The **principle of superposition** tells us what happens if two or more waveforms overlap. This might happen when you are listening to stereo speakers, or when two light beams are focused onto a screen. The result at a particular point is simply the sum of all the disturbances at that point.

Figure 5.4: (a) Constructive interference, (b) destructive interference of two sine waves

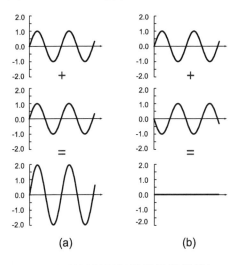

Figure 5.4 shows plots of displacement against **time** at a certain point for two coherent sine waves with the same amplitude. Using the principle of superposition, the lowest graphs show the resultant wave at that point. In both cases, the resultant wave has an amplitude equal to the sum of the amplitudes of the two interfering waves. If the two waves are exactly in phase, as shown in Figure 5.4(a), **constructive interference** occurs, the amplitude of the resultant wave is greater than the amplitude of either individual wave. If they are exactly out-of-phase ('in anti-phase'), the sum of the two disturbances is zero at all times, hence there is no net disturbance. This is called **destructive interference**, as shown in Figure 5.4(b).

The **phase difference** between two waves can be expressed in fractions of a wavelength or as a fraction of a circle, with one whole wavelength being equivalent to a phase difference of 360° or 2π radians. Two waves that are in anti-phase would therefore have a phase difference of $\lambda/2$ or 180° or π radians.

Use the following exercise to investigate the superposition of two coherent waves with phase differences other than 0, π and 2π radians.

The Phase Angle Equation can be used to calculate the phase difference between two points on a single wave or two separate waves.

$$\varphi = 2\pi x/\lambda$$

(5.5)

. . . .

where φ is the phase difference or phase angle measured in radians and x/λ is the fraction of the wavelength.

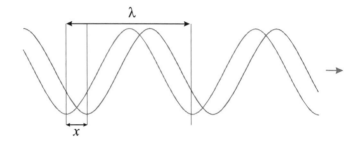

Quiz: Superposition

Go online

Q11: Two sine waves are exactly out of phase at a certain point in space, so they undergo destructive interference. If one wave has amplitude 5.0 cm and the other has amplitude 2.0 cm, what is the amplitude of the resultant disturbance?

a) 0 cm
b) 2.0 cm
c) 2.5 cm
d) 3.0 cm
e) 7.0 cm

. . . .

Q12: A listener is standing midway between two loudspeakers, each broadcasting the same signal in phase. Does the listener hear

a) a loud signal, owing to constructive interference?
b) a quiet signal, owing to destructive interference?
c) a quiet signal, owing to constructive interference?
d) a loud signal, owing to destructive interference?
e) no signal at all?

. . . .

Q13: A radio beacon is transmitting a signal (λ = 200 m) to an aeroplane. When the aeroplane is 4.50 km from the beacon what is the separation between the beacon and the aeroplane in numbers of wavelengths?

a) 0.0225 wavelengths
b) 0.044 wavelengths
c) 22.5 wavelengths
d) 44.0 wavelengths
e) 900 wavelengths

....

Q14: Fourier's theorem tells us that

a) only coherent waves can be added together.
b) any periodic wave is a superposition of harmonic sine and cosine waves.
c) any periodic wave is a superposition of stationary and travelling waves.
d) all sine and cosine waves have the same phase.
e) any periodic wave is made up of a set of sine waves, all with the same amplitude.

....

Q15:

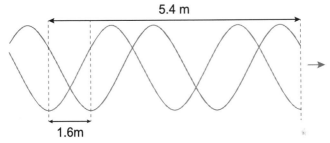

Calculate the phase angle between these two waves.

....

Superposition of two waves

Go online

(a)

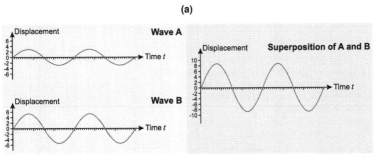

(b)

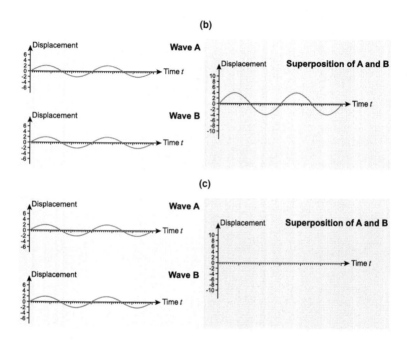

(c)

Q16: Initially the phase difference is zero (a). What do you notice about the size of the amplitude of the resulting wave?

....

Q17: The amplitudes of A and B are set to 2 units (b) and the phase difference to 1 wavelength. What happens?

....

Q18: The phase difference is set to -0.5 wavelengths and the amplitudes of A and B set to 2 units (c). What is the amplitude of the resulting wave?

....

As a practical example of interference, let us look at what happens when coherent waves of identical amplitude are emitted in phase by two loudspeakers. It should be clear that at a point equidistant from each speaker, two waves travelling with the same speed will arrive at exactly the same time. Constructive interference will occur, and the amplitude of the resultant wave will be the sum of the amplitudes of the two individual waves.

Figure 5.5 shows the **wavefronts** from two sources S_1 and S_2, producing waves with identical wavelengths. The wavefronts join points of identical phase, such as the crests of a wave, and the distance between adjacent wavefronts from the same source is equal to λ. Where wavefronts from the two sources overlap (shown by the solid black dots), constructive interference occurs.

Figure 5.5: Interference of waves from two sources

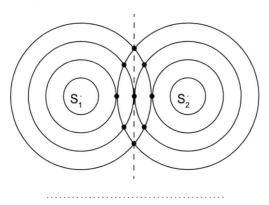

In fact constructive interference will occur at any point where the difference in path length between the waves from each of the two sources is equal to a whole number of wavelengths. At any such point, the arrival of the crest of a wave from the left-hand speaker will coincide with the arrival of a crest from the right-hand speaker, leading to constructive interference.

Put mathematically, the condition for constructive interference of two waves is

$$|l_1 - l_2| = n\lambda$$

(5.6)

where l_1 and l_2 are the distances from source to detector of the two waves, and n is a whole number. (The $||$ around $l_1 - l_2$ means the 'absolute' value, ignoring the minus sign if $l_2 > l_1$.)

If the path difference between the two waves is an odd number of half-wavelengths ($\lambda/2$, $3\lambda/2$, etc.) then destructive interference occurs. The crest of a wave from one speaker will now arrive at the same time as the trough of the wave from the other speaker. If the amplitudes of the two waves are the same, the result of adding them together is zero as in Figure 5.4 (b).

In this case

$$|l_1 - l_2| = \left(n + \frac{1}{2}\right)\lambda$$

(5.7)

....

In terms of phase, we can state that constructive interference occurs when the two waves are in phase, and destructive interference occurs when the two waves are in anti-phase.

Check your understanding of constructive and destructive interference using the superposition shown in Figure 5.5. Suppose S_1 and S_2 are emitting coherent sound waves in phase. What would you expect to hear if you walked from S_1 to S_2? (Answer given below the next worked example.)

Example

Two radio transmitters A and B are broadcasting the same signal in phase, at wavelength 750 m. A receiver is at location C, 6.75 km from A and 3.00 km from B. Does the receiver pick up a strong or weak signal?

The distance from A to C is 6750 m, and the wavelength is 750 m, so in wavelengths, the distance A to C is 6750/750 = 9.00 wavelengths.

Similarly, B to C is 3000 m or 3000/750 = 4.00 wavelengths.

So the path difference AC - BC = 5.00 wavelengths, and since this is a whole number of wavelengths, constructive interference will occur and the receiver will pick up a strong signal.

....

Walking from speaker S_1 to S_2 in Figure 5.5, you would hear the sound rising and falling in loudness, as you moved through regions of constructive and destructive interference.

5.4.2 Fourier series

One very important application of the principle of superposition is in the Fourier analysis of a waveform. Fourier's theorem, first proposed in 1807, states that any periodic wave can be represented by a sum of sine and cosine waves, with frequencies which are multiples (harmonics) of the wave in question. If you had been wondering why so much emphasis has been placed on studying sine and cosine waves, it is because any periodic waveform we might encounter is made up of a superposition of sine and cosine waves.

Put mathematically, any wave can be described by a Fourier series

$$y(t) = \frac{a_0}{2} + \sum_{n=1}^{\infty} a_n \cos(n2\pi ft) + \sum_{n=1}^{\infty} b_n \sin(n2\pi ft)$$

(5.8)

. . . .

Fourier analysis is an important technique in many areas of physics and engineering. For example, the response of an electrical circuit to a non-sinusoidal electrical signal can be determined by breaking the signal down into its Fourier components.

How do synthesisers work?

The word synthesiser derives from "synthesis." The most common type used by synthesisers is subtractive synthesis. Starting with a wave with various frequencies the synthesiser subtracts components of the wave until the desired tone is achieved. Frequencies can either be removed or minimised or alternatively emphasised and made more prominent. This can make a completely different sound to the original wave. Sounds can be made to sound like a drum, a trumpet, a crash, boom or bang sound effect or even computer generated speech.

A synthesiser has a range of different "oscillators" or sound tone generators which produce waves of different shapes such as square waves, sine waves, triangular waves etc. By combining these waves it can make complicated sounds and add musical effects like sustain, attack, decay and release to the notes.

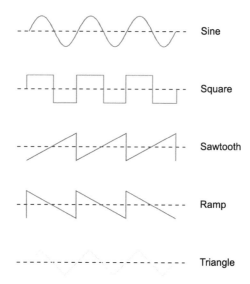

5.5   Stationary waves

We have discussed travelling waves. There is another situation we need to consider, that of stationary or standing waves. Consider a guitar string, or a piece of elastic stretched between two fixed supports. If we pluck the string or elastic, it oscillates at a certain frequency, whilst both ends remain fixed in position (see Figure 5.6).

Figure 5.6: Oscillations of a stretched string held fixed at both ends

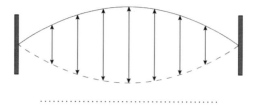

This type of wave is called a **stationary wave**. The name arises from the fact that the waves do not travel along the medium, as we saw in the previous section. Instead, the points of maximum and zero oscillation are fixed. Many different wave patterns are allowed, as will be discussed shortly.

At the fixed ends of the medium (the guitar string, the elastic band or whatever), no oscillations occur. These points are called **nodes**. The points of maximum amplitude oscillations are called **antinodes**. If you have trouble remembering which is which, the nodes are the points where there is 'node-disturbance'.

Stationary Waves

Because of the condition of having a node at each end, we can build up a picture of the allowed modes of oscillation, as shown.

Go online

Figure 5.7: First four harmonics of a transverse standing wave

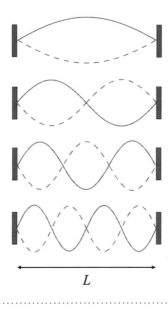

The wavelength of the oscillations is twice the distance between adjacent nodes, so the longest possible wavelength in Figure 5.7 is $\lambda_1 = 2L$. This is called the fundamental mode, and oscillates at the fundamental frequency $f_1 = {}^v/_{\lambda_1} = {}^v/_{2L}$.

Looking at Figure 5.7, we can see that the allowed wavelengths are given by

$$\lambda_1 = 2L, \ \lambda_2 = {}^{2L}/_2, \ \lambda_3 = {}^{2L}/_3 .....$$

$$\lambda_n = {}^{2L}/_n = {}^{\lambda_1}/_n$$

(5.9)

....

The allowed frequencies are therefore given by

$$f_1 = {v}/{2L}, \ f_2 = {2v}/{2L}, \ f_3 = {3v}/{2L}\cdots$$

$$f_n = {nv}/{2L} = nf_1$$

(5.10)

....

These different frequencies are called the **harmonics** of the system, and f_1 is the first harmonic (also called the fundamental mode, as stated earlier). f_2 is called the second harmonic (or the first overtone), f_3 the third harmonic (or second overtone) and so on.

The equation for travelling waves does **not** also describe the motion of stationary waves. It can be proved mathematically that the equation for stationary waves is the superposition of two travelling waves of equal amplitude travelling in opposite directions.

When working out the equation for travelling waves, we stated that every small portion of the medium was performing SHM slightly out of phase with its neighbours, but with the same amplitude. An important difference between stationary and travelling waves is that for stationary waves, each portion of the medium between nodes oscillates **in phase** with its neighbours, but with slightly **different** amplitude.

As with the travelling waves, stationary waves can also be set up for longitudinal as well as transverse waves. Equation 5.9 and Equation 5.10 equally apply for longitudinal and transverse stationary waves.

Example

An organ pipe has length $L = 2.00$ m and is open at both ends. The fundamental stationary sound wave in the pipe has an antinode at each end, and a node in the centre. Calculate the wavelength and frequency of the fundamental note produced. (*Take the speed of sound in air v = 340 m s⁻¹.*)

The distance between two antinodes (like the distance between two nodes) is equal to $\lambda/2$.

$$\therefore \frac{\lambda}{2} = 2.00$$
$$\therefore \lambda = 4.00 \, \text{m}$$

To calculate the frequency of the fundamental, use $n = 1$ in the equation

$$f_n = \frac{nv}{2L}$$
$$\therefore f = \frac{340}{2 \times 2.00}$$
$$\therefore f = \frac{340}{4.00}$$
$$\therefore f = 85.0\,\text{Hz}$$

We could, of course, have calculated f using $f = {v}/{\lambda}$ which gives the same answer.

. . . .

Longitudinal stationary waves

When someone blows across the top of a bottle, a sound wave is heard. This is because a stationary wave has been set up in the air in the bottle. The oscillating air molecules form stationary longitudinal waves. In this exercise you will work out the wavelength of different longitudinal stationary waves.

a) Consider a pipe of length L, with one end open and the other closed. The stationary waves formed by this system have an antinode at the open end and a node at the closed end.

    i   Sketch the fundamental wave in the pipe and calculate its wavelength.

    ii Sketch the next two harmonics.

b) If the pipe is open at both ends, the air molecules are free to oscillate at either end, so there will be an antinode at each end.

    i   For a pipe of length L which is open at both ends, sketch the fundamental wave and calculate its wavelength.

    ii Sketch the next two harmonics.

. . . .

5.6 Beats

Beats are the regular loud and quiet sounds heard when two sound waves of very similar frequencies interfere with one another. e.g. a note of 350Hz and a note of 355Hz would display this effect. The beat frequency is the frequency of the changes between loud and quiet sounds heard by the listener. So if you hear 3 complete cycles of loud and quiet sounds every second, the beat frequency is 3 Hz. Humans can hear beat frequencies of less than or equal to 7Hz. The difference between the frequencies of two notes that interfere to produce the beats is equal to the **beat frequency**.

For example, if a tuning fork plays middle C which is 262Hz and a singer makes a continuous sound of 266Hz then a beat frequency of 4Hz will be heard.

It works because the waves are slightly out of phase and this results in areas of constructive interference creating louder sounds and destructive interference creating quieter sounds. See Figure 5.8 for an visual explanation of how it works.

Figure 5.8: How beats are formed through constructive and destructive interference.

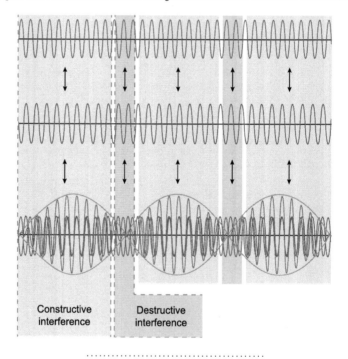

Piano tuners frequently use beats to tune pianos. They open the piano up and pluck the piano strings inside one at a time at the same time as tapping a tuning fork. If no beats can be heard the piano string is in tune as no interference takes place and hence no beat frequency. If beats can be heard the piano tuner tightens or loosens the string until no beats can be heard, if the beat frequency increases, the piano tuner will do the opposite to get nearer to 0Hz and reduce the beat frequency. The piano tuner needs a full set of tuning forks for all the notes on the piano.

Quiz: Stationary waves

Go online

Useful data:

Speed of sound in air	$340\ m\ s^{-1}$

Q19: The third harmonic of a plucked string has frequency 429 Hz. What is the frequency of the fundamental?

a) 1.30 Hz
b) 47.7 Hz
c) 143 Hz
d) 429 Hz
e) 1290 Hz

....

Q20: Which *one* of the following statements is true?

a) Large amplitude oscillations occur at the nodes of a stationary wave.
b) Every point between adjacent nodes of a stationary wave oscillates in phase.
c) The amplitude of the stationary wave oscillations of a plucked string is equal at all points along the string.
d) The distance between adjacent nodes is equal to the wavelength of the stationary wave.
e) Stationary waves only occur for transverse, not longitudinal waves.

....

Q21: A string is stretched between two clamps placed 1.50 m apart. What is the wavelength of the fundamental note produced when the string is plucked?

a) 0.67 m
b) 0.75 m
c) 1.50 m
d) 2.25 m
e) 3.00 m

....

Q22: The fourth harmonic of a standing wave is produced on a stretched string 2.50 m long. How many antinodes are there?

a) 2.5
b) 3
c) 4
d) 5
e) 10

. . . .

Q23: A stationary sound wave is set up in an open pipe 1.25 m long. What is the frequency of the third harmonic note in the pipe?

a) 182 Hz
b) 264 Hz
c) 340 Hz
d) 408 Hz
e) 816 Hz

. . . .

Q24: A guitar string with a frequency of 220 Hz is plucked at the same time as a tuning fork with a frequency of 215 Hz.
How many beats will be heard over a period of 5 seconds?

a) 21 beats
b) 25 beats
c) 26 beats
d) 28 beats
e) 29 beats

. . . .

5.7   Extended information

Web links

There are web links available online exploring the subject further.

. . . .

5.8   Summary

Summary

You should now be able to:

- use all the terms commonly employed to describe waves;

- derive an equation describing travelling sine waves, and solve problems using this equation;

- show an understanding of the difference between travelling and stationary waves;

- calculate the harmonics of a number of stationary wave systems;

> Summary continued
>
> - apply the principle of superposition;
> - be able to explain how beats can be used to tune musical instruments;
> - to understand the term phase difference and use the phase angle equation.

5.9 Assessment

End of topic 5 test

The following test contains questions covering the work from this topic.

Go online

The following data should be used when required:

Speed of light in a vacuum c	$3.00 \times 10^8 m\ s^{-1}$
Speed of sound	$340\ m\ s^{-1}$
Acceleration due to gravity g	$9.8\ m\ s^{-2}$

Q25: A laser produces a monochromatic (single wavelength) beam of light with wavelength 542 nm.

Calculate the frequency of the light.

_____ Hz

. . . .

Q26: A beam of red light (λ = 660 nm) is focused onto a detector which measures a light irradiance of 1.63×10^{-6} W m^{-2}.

Calculate the measured irradiance when the amplitude of the waves is doubled.

_____ W m^{-2}

. . . .

Q27: Suppose a knot is tied in a horizontal piece of string. A train of transverse vertical sine waves are sent along the string, with amplitude 8.5 cm and frequency 1.6 Hz.

Calculate the total distance through which the knot moves in 5.0 s.

_____ cm

. . . .

Q28: A travelling wave is represented by the equation

$$y = 4\sin\left(2\pi\left(1.6t - \frac{x}{5.5}\right)\right)$$

All the quantities in this equation are in SI units.

What is the value of the speed of the wave?

_____ m s^{-1}

. . . .

Q29: A transverse wave travelling along a rope is represented by the equation

$$y = 0.55 \sin\left(2\pi\left(0.35t - \frac{x}{2.5}\right)\right)$$

All the quantities in this equation are in SI units.

Calculate the displacement at the point x = 0.50 m when t = 1.0 s.

_____ m

. . . .

Q30: Two coherent sine waves overlap at a point A. The amplitude of one wave is 6.3 cm, and the amplitude of the other wave is 3.2 cm.

Calculate the **minimum** possible amplitude of the resultant disturbance at A.

_____ cm

. . . .

Q31: Two in-phase speakers A and B are emitting a signal of wavelength 1.04 m. A tape recorder is placed on the straight line between the speakers, 2.69 m from speaker A.

Calculate the **shortest** distance from the tape recorder that speaker B should be placed, to ensure constructive interference where the signal is recorded.

_____ m

. . . .

Q32: Two loudspeakers are emitting a single frequency sound wave in phase. A listener is seated 4.6 m from one speaker and 2.3 m from the other.

Calculate the **minimum** frequency of the sound waves that would allow constructive interference where the listener is seated.

_____ Hz

. . . .

Q33: A guitar string is 0.61 m long.

Calculate the wavelength of the 5th harmonic.

_____ m

. . . .

Q34: A string is stretched between two clamps held 2.25 m apart. The string is made to oscillate at its third harmonic frequency.

Calculate the distance between two adjacent nodes.

_____ m

. . . .

Q35: A plank of wood is placed over a pit 17 m wide. A girl stands on the middle of the plank and starts jumping up and down, jumping upwards from the plank every 1.3 s. The plank oscillates with a large amplitude in its fundamental mode, the maximum amplitude occuring at the centre of the plank.

Calculate the speed of the transverse waves on the plank.

_____ m s^{-1}

. . . .

Q36: A piano is tuned using a 455 Hz tuning fork and a beat frequency of 5 Hz is heard, what was the frequency of the piano string if the piano tuner had to tighten the string to get it into tune?

_____ Hz

. . . .

Topic 6

Interference

Contents

Prerequisite knowledge

- *Refractive index.*
- *Frequency and wavelength.*

Learning objectives

By the end of this topic you should be able to:

- *show an understanding of coherence between light waves;*
- *explain the difference between path length and optical path length, and calculate the latter;*
- *understand what happens when waves reflect off media of higher or lower refractive index.*

6.1   Introduction

Interference of light waves is responsible for the rainbow colours seen on an oil film on a puddle of water, or in light reflected by a soap bubble. In the next few topics we will be looking at the conditions under which these interference effects can take place.

This topic contains some of the background work necessary to fully understand interference. It begins with the concept of coherence, shows the method of calculating the optical path difference between two light rays and finishes with what happens when waves reflect off media of higher or lower refractive index.

6.2   Coherence and optical path difference

The section begins with the concept of coherence and ends with the method of calculating the optical path difference between two light rays.

6.2.1   Coherence

We briefly discussed coherent waves in the 'Introduction to Waves' topic. Two waves are said to be coherent if they have a constant phase relationship. For two waves travelling in air to have a constant phase relationship, they must have the same frequency and wavelength. At any given point, the phase difference between the two waves will be fixed.

It is easier to produce coherent sound waves or microwaves than it is to produce coherent visible electromagnetic waves. Both sound and microwaves can be generated electronically, with loudspeakers or antennae used to emit the waves. The electronic circuits used to generate these waves can 'frequency lock' and 'phase lock' two signals to ensure they remain coherent. In contrast to this, light waves are produced by transitions in individual atoms, and are usually emitted with random phase.

For us to see interference effects, we require two or more sources of coherent light waves. The best source of coherent radiation is a laser, which emits light at a single wavelength, usually in a collimated (non-diverging) beam. Another way to produce coherent light is to split a wave, for example by reflection from a glass slide. Some of the light will be transmitted, the rest will be reflected, and the two parts must be coherent with each other.

Filament light bulbs and strip lights do not emit coherent radiation. Such sources are called extended sources, or incoherent sources. They emit light of many different wavelengths, and light is emitted from every part of the tube or filament.

6.2.2   Optical path difference

In the Section on Superposition in Topic 5, we solved problems in which waves emitted in phase arrived from two different sources. By measuring the paths travelled by both waves, we could determine whether they arrived in phase (interfering constructively) or out of phase (interfering destructively). We are now going to look at a slightly different situation, in which waves emitted in phase by the same source arrive at a detector via different routes.

Figure 6.1: Two light rays travelling along different paths

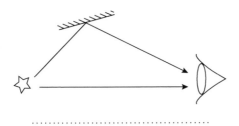

In Figure 6.1 we can see two waves from the same source arriving at the same detector. The situation is similar to that which we saw in the previous Topic. If the **path difference** is a whole number of wavelengths (λ, 2λ, 3λ...) then the two waves will arrive in phase. If the path difference is an odd number of half wavelengths ($\lambda/2$, $3\lambda/2$, $5\lambda/2$...) then the two waves arrive out of phase at the detector and destructive interference takes place.

A further complication can arise if one of the rays passes through a different medium

Figure 6.2: Two light rays travelling different optical path lengths

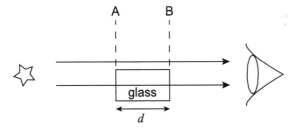

In Figure 6.2, the lower ray passes through a glass block between A and B. An **optical path difference** Δd exists between the two rays, even though they have both travelled the same distance, because the wavelength of the light changes as it travels through the glass. The refractive index of glass, n_{glass}, is greater than the refractive index of air, so the waves travel at a slower speed in the glass. The frequency of the waves does not change, so the wavelength in glass λ_{glass} must be smaller than the wavelength in air, λ. If the refractive index of air is 1.00, then λ_{glass} is given by

$$n_{glass} = \frac{c}{v_{glass}}$$

$$\therefore n_{glass} = \frac{f \times \lambda}{f \times \lambda_{glass}}$$

$$\therefore n_{glass} = \frac{\lambda}{\lambda_{glass}}$$

$$\therefore \lambda_{glass} = \frac{\lambda}{n_{glass}}$$

(6.1)

. . . .

In Figure 6.2, the upper ray is travelling in air all the way from source to detector. The number of wavelengths contained between A and B will be d/λ.

For the lower ray, in the same distance, the number of wavelengths will be

$$\frac{d}{\lambda_{glass}} = \frac{d}{\lambda/n_{glass}} = \frac{d \times n_{glass}}{\lambda}$$

So there will be more wavelengths between A and B in the glass than in the same distance in air. This is shown in Figure 6.3. For a light ray travelling a distance d in a material of refractive index n, the **optical path length** is $n \times d$. To determine the phase difference between two waves travelling between the same source and detector, the **optical path difference** must be a whole number of wavelengths for constructive interference, and an odd number of half-wavelengths for destructive interference. In Figure 6.3, the optical path difference is the difference between the optical path lengths of both waves travelling from A to B. For the upper ray, the optical path length of AB = $n_{air} \times d = d$, since $n_{air} \approx 1.00$. For the lower ray, the optical path length of AB is $n_{glass} \times d$. Thus the optical path difference is $(n_{glass} - n_{air}) \times d$.

Figure 6.3: Optical path difference between air and glass

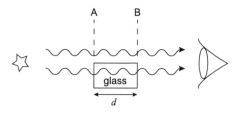

. . . .

In general, for two rays of light travelling the same distance d in media with refractive indices n_1 and n_2, the interference conditions are

constructive interference (waves emerge in phase) $\qquad (n_1 - n_2)\,d = m\lambda$

destructive interference (waves emerge exactly out of phase) $\quad (n_1 - n_2)\,d = \left(m + \frac{1}{2}\right)\lambda$

m is an integer. Note that we will talk about an "optical path difference" even when we are studying radiation from outside the visible part of the electromagnetic spectrum.

Example

Two beams of microwaves with a wavelength of 6.00×10^{-3} m are emitted from a source. One ray of the waves travels through air to a detector 0.050 m away. Another ray travels the same distance through a quartz plate to the detector. Do the waves interfere constructively or destructively at the detector, if the refractive index of quartz is 1.54?

The optical path difference Δd in this case is

$$\begin{aligned}
\Delta d &= (n_{quartz} - n_{air}) \times d \\
\therefore \Delta d &= (1.54 - 1) \times 0.050 \\
\therefore \Delta d &= 0.54 \times 0.050 \\
\therefore \Delta d &= 0.027 \text{ m}
\end{aligned}$$

To find out how many wavelengths this optical path length is, divide by the wavelength in air

$$\frac{0.027}{\lambda} = \frac{0.027}{6.00 \times 10^{-3}} = 4.5 \text{ wavelengths}$$

So the waves arrive at the detector exactly out of phase, and hence interfere destructively.

. . . .

Go online

Quiz: Coherence and optical paths

Q1: Two light waves are coherent if

a) they have the same speed.
b) their amplitudes are identical.
c) the difference in their frequencies is constant.
d) their phase difference is constant.
e) the difference in their wavelengths is constant.

....

Q2: A radio transmitter emits waves of wavelength 500 m, A receiver dish is located 4.5 km from the transmitter. What is the path length, in wavelengths, from the transmitter to the receiver?

a) 0.11 wavelengths
b) 0.5 wavelengths
c) 2 wavelengths
d) 5 wavelengths
e) 9 wavelengths

....

Q3: Light waves of wavelength 450 nm travel 0.120 m through a glass block ($n = 1.50$).

What is the optical path length travelled?

a) 8.10×10^{-9} m
b) 4.00×10^{-4} m
c) 0.180 m
d) 0.800 m
e) 4.00×10^{5} m

....

Q4: Two light rays travel in air from a source to a detector. Both travel the same distance from source to detector, but one ray travels for 2.50 cm of its journey through a medium of refractive index 1.35.

What is the optical path difference between the two rays?

a) 8.75×10^{-3} m
b) 3.38×10^{-3} m
c) 0.0250 m
d) 14.0 m
e) Depends on the wavelength of the light rays.

....

Q5: In a similar set-up to question 4, two microwaves ($\lambda = 1.50$ cm) from the same source arrive with a phase difference of exactly 4.00 wavelengths at the detector. The waves have travelled the same distance, but one has gone through a sheet of clear plastic ($n = 1.88$) while the other has travelled through air all the way.

What is the thickness of the plastic sheet?

a) 0.0319 m
b) 0.0682 m
c) 0.220 m
d) 4.55 m
e) 14.7 m

....

6.3 Reflection of waves

The phase of a wave may be changed when it is reflected. For light waves, we must consider the refractive indices of the two media involved. For example, consider a light wave travelling in air ($n = 1.00$) being reflected by a glass surface ($n = 1.50$). In this case the wave is travelling in a low refractive index medium, and is being reflected by a higher refractive index medium. Whenever this happens, there is a phase change of 180° (π radians). A wave crest becomes a wave trough on reflection. This is shown in Figure 6.4, where we can see a crest (labelled c) before reflection comes back as a trough (t) after reflection.

Figure 6.4: Phase change upon reflection at a higher refractive index material

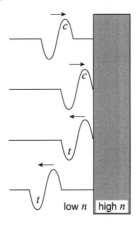

You would see exactly the same effect happening if you sent a wave along a rope that was fixed at one end. As in Figure 6.4, a wave that has a crest leading a trough is reflected back as a wave with a trough leading a crest.

This can be demonstrated effectively by holding a slinky fixed at one end to represent a higher refractive index and sending a transverse wave along the slinky towards the fixed end.

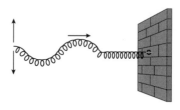

Reflection of a pulse at a fixed end

Go online

There is an activity available online demonstrating a pulse moving through the rope fixed to the wall.

. . . .

There is no phase change when a light wave travelling in a higher index material is reflected at a boundary with a material that has a lower index. If our light wave is travelling in glass, the phase of the wave is not changed when it is reflected at a boundary with air. A wave crest is still a wave crest upon reflection. This reflection is shown in Figure 6.5.

Figure 6.5: No phase change upon reflection at a lower refractive index material

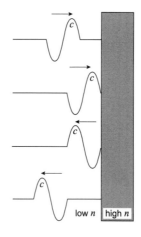

. . . .

A wave sent along a rope that is unsecured at the end would behave in exactly the same way. A wave which travels as a crest leading a trough remains as a wave with the crest leading the trough upon reflection.

Again this can be demonstrated using the slinky but this time having the slinky attached to a piece of string to represent the lower medium.

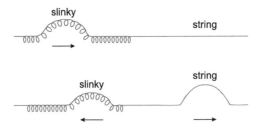

The importance of the phase change in certain reflections is that the optical path of the reflected wave is changed. If the phase of a wave is changed by 180° it is as if the wave has travelled an extra $\lambda/2$ distance compared to a wave whose phase has not been changed by the reflection.

6.4 Extended information

Web links

There are web links available online exploring the subject further.

. . . .

6.5 Summary

Summary

You should now be able to:

- state the condition for two light beams to be coherent;

- explain why the conditions for coherence are more difficult to achieve for light than for sound and microwaves;

- define the term "optical path difference" and relate it to phase difference.

6.6 Assessment

End of topic 6 test

The following test contains questions covering the work from this topic.

Go online

The following data should be used when required:

Speed of light in a vacuum c	$3.00 \times 10^8 \ m \ s^{-1}$
Speed of sound	$340 \ m \ s^{-1}$
Acceleration due to gravity g	$9.8 \ m \ s^{-2}$

Q6: Monochromatic light of wavelength 474 nm travels through a glass lens of thickness 21.5 mm and refractive index 1.52.

Calculate the optical path length through the lens.

_____ mm

. . . .

Q7: Two light rays from a coherent source travel through the same distance to reach a detector. One ray travels 23.7 mm through glass of refractive index 1.52. The other ray travels in air throughout.

Calculate the optical path difference between the rays.

_____ mm

. . . .

Topic 7

Division of amplitude

Contents

Prerequisite knowledge

- *Waves (Topic 5).*
- *Interference (Topic 6).*

Learning objectives

By the end of this topic you should be able to:

- *show an understanding of interference by division of amplitude;*
- *understand how thin film interference works and how it can be used in processes such as creating anti-reflecting coatings;*
- *understand how wedge fringes work and how they can be used to find the thickness of very small objects.*

7.1 Introduction

This topic is all about a process called **interference by division of amplitude**. This is rather a long title for a simple idea. You should already know what we mean by interference from previous Topic 6. Division of amplitude means that each wave is being split, with some of it travelling along one path, and the remainder following a different path. When these two parts are recombined, it is the difference in optical paths taken by the two waves that determines their phase difference, and hence how they interfere when they recombine.

7.2 Thin film interference

One example of interference by division of amplitude with which you are probably familiar is called thin film interference. When oil or petrol is spilt onto a puddle of water, we see a multicoloured film on the surface of the puddle. This is due to the thin film of oil formed on the surface of the puddle. Sunlight is being reflected from the film and the puddle. The film appears multicoloured because of constructive and destructive interference of the sunlight falling on the puddle. In this section we will examine this effect more closely.

Figure 7.1 shows light waves falling on a thin film of oil on the surface of water.

Figure 7.1: Light reflected from a thin film of oil on water

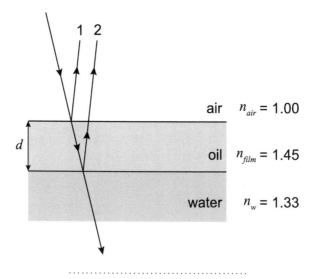

A pine cone thrown into an oily puddle

Photo by David Lee / CC BY-SA 2.0

Light is reflected back upwards from both the air-oil boundary (ray 1) and the oil-water boundary underneath it (ray 2). Someone looking at the reflected light will see the superposition of these two rays. Note that this situation is what we have called interference by division of amplitude. For any light wave falling on the oil film, some of the wave is reflected from the surface of the oil film. Some light is reflected by the water surface, and some is transmitted into the water. The two reflected rays travel different paths before being recombined. Under what conditions will the two rays interfere constructively or destructively?

To keep the analysis simple, we will assume the angle of incidence in Figure 7.1 is $0°$, so that ray 2 travels a total distance of $2d$ in the film. The optical path difference between rays 1 and 2 is therefore $2n_{film}d$. But there is another source of path difference. Ray 1 has undergone a $\lambda/2$ phase change, since it has been reflected at a higher refractive index medium.

The total optical path difference is therefore equal to

$$2n_{film}d + \frac{\lambda}{2}$$

For constructive interference, this optical path difference must equal a whole number of wavelengths.

$$2n_{film}d + \frac{\lambda}{2} = m\lambda$$

where m = 1, 2, 3...

$$2n_{film}d = m\lambda - \frac{\lambda}{2}$$
$$\therefore 2n_{film}d = \left(m - \frac{1}{2}\right)\lambda$$
$$\therefore d = \frac{\left(m - \frac{1}{2}\right)\lambda}{2n_{film}}$$

(7.1)

. . . .

Equation 7.1 gives us an expression for the values of film thickness d for which reflected light of wavelength λ will undergo constructive interference, and hence be reflected strongly.

For destructive interference, the optical path difference must equal an odd number of half-wavelengths

$$2n_{film}d + \frac{\lambda}{2} = \left(m + \frac{1}{2}\right)\lambda$$

where $m = 1, 2, 3...$

$$\therefore 2n_{film}d = \left(m + \frac{1}{2}\right)\lambda - \frac{\lambda}{2}$$
$$\therefore 2n_{film}d = m\lambda$$
$$\therefore d = \frac{m\lambda}{2n_{film}}$$

(7.2)

. . . .

Equation 7.2 tells us for which values of d reflected light of wavelength λ will undergo destructive interference, and hence be reflected weakly.

Example

Using Figure 7.1, what is the *minimum* thickness of oil film which would result in destructive interference of green light (λ = 525 nm) falling on the film with angle of incidence 0°?

For destructive interference we will use Equation 7.2, with λ = 525 nm = 5.25×10^{-7} m. For the minimum film thickness, we set m equal to 1.

$$d = \frac{m\lambda}{2n_{film}}$$
$$\therefore d = \frac{1 \times 5.25 \times 10^{-7}}{2 \times 1.45}$$
$$\therefore d = \frac{1 \times 5.25 \times 10^{-7}}{2.90}$$
$$\therefore d = 1.81 \times 10^{-7} \text{ m}$$

. . . .

If we have sunlight falling on an oil film, then we have a range of wavelengths present. Also, an oil film is unlikely to have the same thickness all along its surface. So we will have certain wavelengths interfering constructively where the film has one thickness, whilst the same wavelengths may be interfering destructively at a part of the surface where the film thickness is different. The overall effect is that the film appears to be multi-coloured.

Thin film interference

At this stage there is an online activity showing interference as the thickness of a thin film is varied.

Go online

. . . .

An important application of thin film interference is anti-reflection coatings used on camera lenses. A thin layer of a transparent material such as magnesium fluoride (n = 1.38) is deposited on a glass lens (n = 1.50), as shown in Figure 7.2.

Figure 7.2: Anti-reflection coating deposited on a glass lens

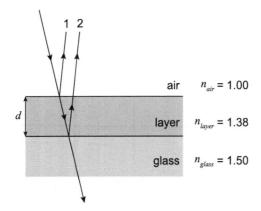

....

Once again, if we consider the light falling with an angle of incidence of $0°$ on the lens, then the optical path difference between rays 1 and 2 is $2 \times n_{coating} \times d$. Both rays are reflected by media with a greater value of refractive index, so they **both** undergo the same $\lambda/2$ phase change on reflection and there is no extra optical path difference as there was with the oil film on water. So for destructive interference

$$2n_{coating}d = \left(m + \frac{1}{2} \right) \lambda$$
$$\therefore d = \frac{\left(m + \frac{1}{2} \right) \lambda}{2n_{coating}}$$

(7.3)

....

In this equation, m = 0, 1, 2...

The minimum coating thickness that will result in destructive interference is given by putting m = 0 in Equation 7.3:

$$d = \frac{\left(0 + \frac{1}{2} \right) \lambda}{2n_{coating}}$$
$$\therefore d = \frac{\lambda}{4n_{coating}}$$

(7.4)

....

So a coated **(bloomed) lens** can be made non-reflecting for a specific wavelength of light, and Equation 7.4 gives the minimum coating thickness for that wavelength.

Example

For the coated lens shown in Figure 7.2, what is the minimum thickness of magnesium fluoride that can be used to make the lens non-reflecting at $\lambda = 520$ nm?

Use Equation 7.4

$$d = \frac{\lambda}{4n_{coating}}$$
$$\therefore d = \frac{5.20 \times 10^{-7}}{4 \times 1.38}$$
$$\therefore d = 9.42 \times 10^{-8} \text{ m}$$

. . . .

Since most lenses are made to operate in sunlight, the thickness of the coating is designed to produce destructive interference in the centre of the visible spectrum. The previous example produced an anti-reflection coating in the green part of the spectrum, which is typical of commercial coatings. The extremes of the visible spectrum - red and violet - do not undergo destructive interference upon reflection, so a coated lens often looks reddish-purple under everyday lighting.

Quiz: Thin film interference

Go online

 Useful data:

Refractive index of air	*1.00*
Refractive index of water	*1.33*

Q1: Two coherent light rays emitted from the same source will interfere constructively if

a) they undergo a phase change on reflection.
b) they travel in materials with different refractive indices.
c) their optical path difference is an integer number of wavelengths.
d) their optical path difference is an odd number of half-wavelengths.
e) they have different wavelengths.

. . . .

Q2: What is the minimum thickness of oil film ($n = 1.48$) on water that will produce destructive interference of a beam of light of wavelength 620 nm?

a) 1.05×10^{-7} m
b) 2.09×10^{-7} m
c) 2.33×10^{-7} m
d) 3.14×10^{-7} m
e) 2.07×10^{-6} m

....

Q3: A soap film, with air on either side, is illuminated by electromagnetic radiation normal to its surface. The film is 2.00×10^{-7} m thick, and has refractive index 1.40. Which wavelengths will be intensified in the reflected beam?

a)   200 nm and 100 nm
b)   560 nm and 280 nm
c)   1120 nm and 373 nm
d)   1120 nm and 560 nm
e)   2240 nm and 747 nm

....

Q4: Why do lenses coated with an anti-reflection layer often appear purple in colour when viewed in white light?

a)   You cannot make an anti-reflection coating to cut out red or violet.
b)   The coating is too thick to work properly.
c)   The green light is coherent but the red and violet light is not.
d)   The coating is only anti-reflecting for the green part of the visible spectrum.
e)   Light only undergoes a phase change upon reflection at that wavelength.

....

Q5: What is the minimum thickness of magnesium fluoride ($n = 1.38$) that can form an anti-reflection coating on a glass lens for light with wavelength 500 nm?

a) 9.06×10^{-8} m
b) 1.36×10^{-7} m
c) 1.81×10^{-7} m
d) 2.72×10^{-7} m
e) 5.00×10^{-7} m

....

There is one final point to add before we leave thin film interference. In the previous Topic, we discussed coherence, and stated that we required coherent radiation to observe interference effects. But we have seen that we can produce interference with an incoherent source such as sunlight. How is this possible? The answer is that the process of interference by division of amplitude means that we are taking one wave, splitting it up and then recombining it with itself. We do not have one wave interfering with another, and so we do not need a source of coherent waves. Division of amplitude means we can use an extended source to produce interference.

7.3 Wedge fringes

Interference fringes can also be produced by a thin wedge of air. If we place a thin piece of foil between two microscope slides at one end, we form a thin air wedge. This is shown in Figure 7.3, where the size of the wedge angle has been exaggerated for clarity.

Figure 7.3: The side and top views of a wedge fringes experiment

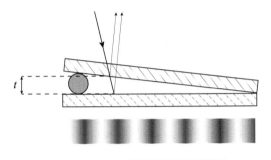

When the wedge is illuminated with a monochromatic source, bright and dark bands are seen in the reflected beam. Interference is taking place between rays reflected from the lower surface of the top slide and the upper surface of the bottom slide. Because of the piece of foil, the thickness of the air wedge is increasing from right to left, so the optical path difference between the reflected rays is increasing. A bright fringe is seen when the optical path difference leads to constructive interference, and a dark fringe occurs where destructive interference is taking place.

We can calculate the wedge thickness required to produce a bright or dark fringe. We will be considering light falling normally onto the glass slides. Remember that the angle between the slides is extremely small, so it can be assumed to be approximately zero in this analysis. The path difference between the two rays in Figure 7.3 is the extra distance travelled **in air** by the ray reflected from the lower slide, which is equal to $2t$.

Do either of the rays undergo a phase change upon reflection?

The answer is yes - the ray reflected by the lower slide is travelling in a low refractive index medium (air) and being reflected at a boundary with a higher n medium (glass) so this ray does undergo a $\lambda/2$ phase change. The other ray does not, as it is travelling in the higher n medium. So the total optical path difference between the two rays is

$$2t + \frac{\lambda}{2}$$

For constructive interference and a bright fringe,

$$2t + \frac{\lambda}{2} = m\lambda$$

$$\therefore 2t = \left(m - \frac{1}{2}\right)\lambda$$

$$\therefore t = \frac{\left(m - \frac{1}{2}\right)\lambda}{2}$$

(7.5)

. . . .

In this case, $m = 1, 2, 3...$

For destructive interference, leading to a dark fringe

$$2t + \frac{\lambda}{2} = \left(m + \frac{1}{2}\right)\lambda$$

$$\therefore 2t = m\lambda$$

$$\therefore t = \frac{m\lambda}{2}$$

(7.6)

. . . .

Here, m is a whole number again, but we can also put $m = 0$, which corresponds to $t = 0$. Check back to Figure 7.3, and you can see that the $m = 0$ case gives us a dark fringe where the two slides are touching and $t = 0$.

Finally, the fringe separation can be determined if we know the size and separation of the glass slides.

Figure 7.4: Wedge separation analysis

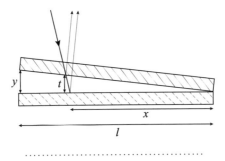

The length of the glass slides shown in Figure 7.4 is l m, and the slides are separated by y m at one end. The mth dark fringe is formed a distance x m from the end where the slides are in contact. By looking at similar triangles in Figure 7.4, we can see that

$$\frac{t}{x} = \frac{y}{l}$$
$$\therefore x = \frac{tl}{y}$$

Substituting for t using Equation 7.6

$$x = \frac{m\lambda l}{2y}$$

The distance Δx between the mth and $(m+1)^{th}$ dark fringes (the fringe separation) is therefore

$$\Delta x = \frac{(m+1)\lambda l}{2y} - \frac{m\lambda l}{2y}$$
$$\therefore \Delta x = \frac{\lambda l}{2y}$$

(7.7)

Go online

Wedge fringes demonstration

There is an online activity which demonstrates how varying the thickness of the gap, colour of the light and the index of refraction of the material between the slides (normally air).

. . . .

Wedge fringes

Go online

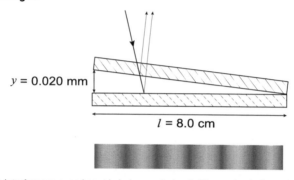

A wedge interference experiment is being carried out. The wedge is formed using two microscope slides, each of length 8.0 cm, touching at one end. At the other end, the slides are separated by a 0.020 mm thick piece of foil. What is the fringe spacing when the experiment is carried out using:

1. A helium-neon laser, at wavelength 633 nm?

2. An argon laser, at wavelength 512 nm?

. . . .

Go online

Quiz: Wedge fringes

Q6: In a wedge interference experiment carried out in air using monochromatic light, the 8th bright fringe occurs when the wedge thickness is 1.80×10^{-6} m.

What is the wavelength of the light?

a) 225 nm
b) 240 nm
c) 450 nm
d) 480 nm
e) 960 nm

. . . .

Q7: An air-wedge interference experiment is being carried out using a mixture of blue light (λ = 420 nm) and red light (λ = 640 nm).

What is the colour of the bright fringe that is seen closest to where the glass slides touch?

a) Blue
b) Red
c) Purple - both blue and red appear at the same place
d) It is dark, as the two colours cancel each other out
e) Impossible to say without knowing the thickness of the wedge

. . . .

Q8: A wedge interference experiment is carried out using two glass slides 10.0 cm long, separated at one end by 1.00×10^{-5} m.

What is the fringe separation when the slides are illuminated by light of wavelength 630 nm?

a) 1.00×10^{-3} m
b) 1.25×10^{-3} m
c) 1.58×10^{-3} m
d) 3.15×10^{-3} m
e) 6.30×10^{-3} m

. . . .

Q9: What is the minimum air-wedge thickness that would produce a bright fringe for red light of wavelength 648 nm?

a) 81.0 nm
b) 162 nm
c) 324 nm
d) 648 nm
e) 972 nm

. . . .

Q10: During a wedge fringes experiment with monochromatic light, air between the slides is replaced by water.

What would happen to the fringes?

a) The bright fringes would appear brighter.
b) The fringes would move closer together.
c) The fringes would move further apart.
d) There would be no difference in their appearance.
e) The colour of the fringes would change.

. . . .

7.4   Extended information

Web links

There are web links available online exploring the subject further.

....

7.5   Summary

Interference of light waves can be observed when two (or more) coherent beams are superposed. This usually requires a source of coherent light waves. Division of amplitude - splitting a wave into two parts which are later re-combined - can be used to produce interference effects without requiring a coherent source.

Two experimental arrangements for viewing interference by division of amplitude have been presented in this topic. In thin film interference, one ray travels an extra distance through a different medium. The thickness and refractive index of the film must be known in order to predict whether constructive or destructive interference takes place. In a thin wedge interference experiment, only the thickness of the wedge is required.

Summary

You should now be able to:

- state what is meant by the principle of division of amplitude, and describe how the division of amplitude allows interference to be observed using an extended source;

- state the conditions under which a light wave will undergo a phase change upon reflection;

- derive expressions for maxima and minima to be formed in a "thin film" reflection, and perform calculations using these expressions;

- explain the formation of coloured fringes when a thin film is illuminated by white light;

- explain how a lens can be made non-reflecting for a particular wavelength of light;

- derive the expression for the minimum thickness of a non-reflecting coating, and carry out calculations using this expression;

- explain why a coated ("bloomed") lens appears coloured when viewed in daylight;

- derive the expression for the distance between fringes formed by "thin wedge" reflection, and carry out calculations using this expression.

7.6 Assessment

End of topic 7 test

The following test contains questions covering the work from this topic.

Go online

The following data should be used when required:

Speed of light in a vacuum c	$3.00 \times 10^8 m\ s^{-1}$
Speed of sound	$340\ m\ s^{-1}$
Acceleration due to gravity g	$9.8\ m\ s^{-2}$

Q11: A soap bubble consists of a thin film of soapy water ($n = 1.32$) surrounded on both sides by air.

Calculate the minimum film thickness if light of wavelength 602 nm is strongly reflected by the film.

_____ m

....

Q12: White light is incident on an oil film (thickness 2.66×10^{-7} m, refractive index 1.35) floating on water (refractive index 1.33). The white light has a wavelength range of 350 - 750 nm.

Calculate the two wavelengths in this range which undergo destructive interference upon reflection.

1. Calculate the longer wavelength.

   _____ nm

2. Calculate the shorter wavelength.

   _____ nm

....

Q13: A 1.02×10^{-7} m coating of a transparent polymer ($n = 1.33$) is deposited on a glass lens to make an anti-reflection coating.

If the lens has refractive index 1.52, calculate the wavelength of light in the range 350 - 750 nm for which the coating is anti-reflecting.

_____ nm

....

Q14: In a standard air wedge interference experiment, calculate the thickness of the wedge which gives the 10th bright fringe for light of wavelength 474 nm.

_____ nm

....

Q15: Wedge fringes are formed in the air gap between two glass slides of length 16.5 cm, separated at one end by a 10.0 μm piece of paper. The wedge is illuminated by monochromatic light of wavelength 510 nm.

Calculate the distance between adjacent dark fringes. Give your answer in metres.

_____ m

. . . .

Q16: Two glass slides are laid together, separated at one end by a 15.0 μm sliver of foil. The slides each have length 10.0 cm.

When illuminated by monochromatic light, a series of light and dark reflection fringes appear, with fringe separation 1.75 mm.

Calculate the wavelength of the light.

_____ nm

. . . .

Topic 8

Division of wavefront

Contents

Prerequisite knowledge

- *A prior learning for this topic is an understanding of coherent light waves.*

Learning objectives

By the end of this topic you should be able to:

- *show an understanding of the difference between interference by division of wavefront and division of amplitude;*

- *describe the Young's slits experiment;*

- *derive the expression for fringe spacing in a Young's slits experiment, and use this expression to determine the wavelength of a monochromatic source.*

8.1   Introduction

The work in this topic will concentrate on the experiment known as Young's slits, which is an example of interference by division of wavefront. We will see the difference between this process and interference by division of amplitude.

The experimental arrangement for producing interference will be described, followed by an analysis of the interference pattern. This analysis will reveal that the spacing between interference fringes is in direct proportion to the wavelength of the light.

8.2   Interference by division of wavefront

Light emitted from a point source radiates uniformly in all directions. This is often illustrated by showing the wavefronts perpendicular to the direction of travel, as in Figure 8.1.

Figure 8.1: Wavefronts emitted by a point source

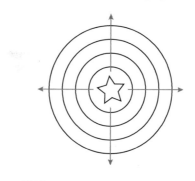

Each wavefront joins points in phase, for example the crest or trough of a wave. If the light source is monochromatic then the points on the wavefront are coherent, as they have the same wavelength and are in phase. If we can take two parts of a wavefront and combine them, then we will see interference effects. This is known as **interference by division of wavefront**. This is a different process to division of amplitude, in which a single wave was divided and re-combined. Here we are combining two separate waves. The two waves must be coherent to produce stable interference when they are combined.

An extended source acts like a collection of point sources, and cannot be used for a division of wavefront experiment. To overcome this problem, an extended source is often used behind a small aperture in a screen (see Figure 8.2). The size of the aperture must be of the same order of magnitude as the wavelength of the light. In this way the light appears to come from a point source.

....

In the next Section, we will see how coherent waves produced in this manner can be combined to show interference effects. Note that if a coherent source such as a laser is used, we do not need to pass the beam through an aperture.

8.3 Young's slits experiment

This experiment is one of the earliest examples of interference by division of wavefront, first carried out by Thomas Young in 1801. The experimental arrangement is shown in Figure 8.3.

Figure 8.3: Young's slits experiment

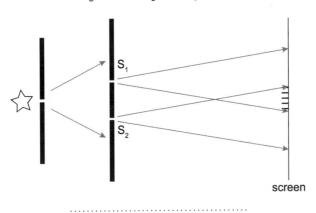

....

Monochromatic light is passed through one narrow slit to give a coherent source, as described earlier. The division of wavefront takes place at the two slits S_1 and S_2. These slits are typically less than 1 mm apart, and act as point sources. A screen is placed about 1 m from the slits. Where the two beams overlap, a symmetrical pattern of fringes is formed, with a bright fringe at the centre.

Figure 8.4: Young's slits interference fringes

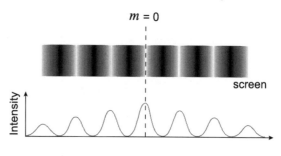

Figure 8.4 shows the fringe pattern seen on the screen. The lower part of Figure 8.4 shows a plot of irradiance across the screen. The brightest fringe occurs at the centre of the interference pattern as this point is equidistant from the slits, and so the waves arrive exactly in phase. The first dark fringes occur on either side of this when the optical path difference between the beams is exactly half a wavelength. This is followed by the next bright fringe, due to a path difference of exactly one wavelength, and so on.

In general, a bright fringe occurs when the path difference between the two beams is $m\lambda$, where $m = 0, 1, 2...$ The central bright fringe corresponds to $m = 0$.

The fringe spacing can be analysed to determine the wavelength of the light.

Figure 8.5: Analysis of Young's slits experiment

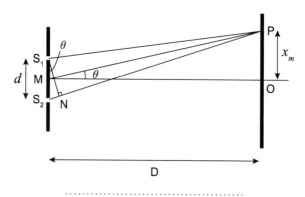

In Figure 8.5, the slits are separated by a distance d, and M marks the midpoint between the slits S_1 and S_2. The screen is placed a distance D ($>> d$) from the slits, with O being the point directly opposite M where the $m = 0$ bright fringe occurs. The mth bright fringe is located at the point P, a distance

$$x_m$$

from O.

The path difference between the two beams is $S_2P - S_1P = m\lambda$. If the length PN equal to S_1P is marked on S_2P, then the path difference is the distance $S_2N = m\lambda$.

Since PM is very much larger than S_1S_2, the line S_1N meets S_2P at approximately a right angle and S_1S_2N is a right-angled triangle. In this triangle

$$\sin\theta = \frac{S_2N}{S_1S_2} = \frac{m\lambda}{d}$$

We can also look at the right-angled triangle formed by $M\ P\ O$

$$\tan\theta = \frac{OP}{MO} = \frac{x_m}{D}$$

Because θ is very small, $\sin\theta \approx \tan\theta \approx \theta$. Therefore

$$\frac{x_m}{D} = \frac{\lambda}{d}$$
$$\therefore x_m = \frac{m\lambda D}{d}$$

To find the separation Δx between fringes, we need to find the distance $x_{m+1} - x_m$ between the $(m+1)^{th}$ and m^{th} bright fringes

$$\Delta x = x_{m+1} - x_m$$
$$\therefore \Delta x = \frac{(m+1)\lambda D}{d} - \frac{m\lambda D}{d}$$
$$\therefore \Delta x = \frac{\lambda D}{d}$$

(8.1)

....

Rearranging Equation 8.1 in terms of λ

$$\lambda = \frac{\Delta x d}{D}$$

(8.2)

. . . .

So a Young's slits experiment can be used to determine the wavelength of a monochromatic light source.

Carry out the following activity to see how the experimental parameters affect the appearance of the interference fringes.

Young's slits experiment

Go online

At this stage there is an online activity which demonstrate Young's slits experiment where the wavelength, slit separation and screen location can be changed.

. . . .

Example

A Young's slits experiment is set up with a slit separation of 0.400 mm. The fringes are viewed on a screen placed 1.00 m from the slits. The separation between the $m = 0$ and $m = 10$ bright fringes is 1.40 cm. What is the wavelength of the monochromatic light used?

We are given the separation for 10 fringes as 1.40 cm, so the fringe separation $\Delta x = 0.140$ cm. Converting all the distances involved into metres, we have $\Delta x = 1.40 \times 10^{-3}$ m, $d = 4.00 \times 10^{-4}$ m and $D = 1.00$ m. Using Equation 8.2

$$\lambda = \frac{\Delta x d}{D}$$
$$\therefore \lambda = \frac{1.40 \times 10^{-3} \times 4.00 \times 10^{-4}}{1.00}$$
$$\therefore \lambda = 5.60 \times 10^{-7} \text{m}$$
$$\therefore \lambda = 560 \text{ nm}$$

. . . .

There is one more problem to consider - what happens if a Young's slits experiment is performed with white light? What does the interference pattern look like then?

If white light is used, there is no difference to the $m = 0$ bright fringe. Since this appears at an equal distance from both slits, then constructive interference will occur whatever the wavelength. For the $m = 1, 2, 3...$ fringes, Equation 8.1 tells us that the fringe separation $\Delta x \propto \lambda$, so the fringe separation is smallest for short wavelengths. Thus the violet end of the spectrum produces the $m = 1$ bright fringe closest to $m = 0$. The red end of the spectrum produces a bright fringe at a larger separation. At higher orders, the fringe patterns of different colours will overlap.

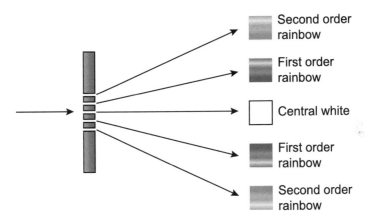

Second order rainbow

First order rainbow

Central white

First order rainbow

Second order rainbow

Quiz: Young's slits

Q1:   The Young's slits experiment is an example of

Go online

a)   interference by division of amplitude.
b)   interference by division of wavefront.
c)   the Doppler effect.
d)   phase change.
e)   refractive index.

Q2:   A Young's slits experiment is set up using slits 0.50 mm apart. The fringes are detected 1.2 m from the slits, and are found to have a fringe spacing of 0.80 mm.

What is the wavelength of the radiation used?

a)   113 nm
b)   128 nm
c)   200 nm
d)   333 nm
e)   720 nm

Q3: What happens to the fringes in a Young's slits experiment if a red light source (λ = 640 nm) is replaced by a green light source (λ = 510 nm)?

a) The fringes disappear.
b) The fringes move closer together.
c) The fringes move further apart.
d) There is no change to the position of the fringes.
e) No fringes are formed in either case.

....

Q4: A Young's slits experiment is carried out by passing coherent light of wavelength 480 nm through slits 0.50 mm apart, with a screen placed 2.5 m from the slits.

What is the spacing between bright fringes viewed on the screen?

a) 2.4 mm
b) 3.8 mm
c) 4.8 mm
d) 5.8 mm
e) 9.6 mm

....

Q5: Using a similar set-up to the previous question, with a red light source at 650 nm, how far from the central bright fringe is the 5th bright fringe formed?

a) 0.00065 m
b) 0.0033 m
c) 0.0041 m
d) 0.0065 m
e) 0.016 m

....

8.4   Extended information

Web links

There are web links available online exploring the subject further.

....

8.5 Summary

Summary

You should now be able to:

- show an understanding of the difference between interference by division of wavefront and division of amplitude;

- describe the Young's slits experiment;

- derive the expression for fringe spacing in a Young's slits experiment, and use this expression to determine the wavelength of a monochromatic source.

8.6 Assessment

End of topic 8 test

The following test contains questions covering the work from this topic.

Go online

 The following data should be used when required:

Acceleration due to gravity g	$9.8\ m\ s^{-2}$
Speed of sound	$340\ m\ s^{-1}$
Speed of light in a vacuum c	$9.8\ m\ s^{-2}$

Q6: A Young's slits experiment is carried out using monochromatic light of wavelength 525 nm. The slit separation is 4.35×10^{-4} m, and the screen is placed 1.00 m from the slits.

Calculate the spacing between adjacent bright fringes.

---------- m

. . . .

Q7: The wavelength of a monochromatic light source is determined by a Young's slits experiment. The slits are separated by 4.95×10^{-4} m and located 2.50 m from the viewing screen.

If the fringe spacing is 2.48 mm, calculate the wavelength of the light.

---------- nm

. . . .

Q8: In a Young's slits experiment, the fringes are viewed on a screen 1.45 m from the slits. The 8th bright fringe is observed 12.8 mm from the central maximum. The slit separation is 0.400 mm.

Calculate the wavelength of the monochromatic light source.

_____ nm

. . . .

Q9: Coherent light at wavelength 490 nm is passed through a pair of slits placed 1.65 mm apart. The resulting fringes are viewed on a screen placed 1.50 m from the slits.

Calculate how far from the central bright fringe the 5th bright fringe appears.

_____ mm

. . . .

Q10: A Young's slits experiment was carried out using two monochromatic sources A and B. The longer wavelength source (A) had wavelength 642 nm. It was found that the 5th bright fringe from A occurred at the same distance from the central bright fringe as the 6th bright fringe from B.

Calculate the wavelength of B.

_____ nm

. . . .

Topic 9

Polarisation

Contents

Prerequisite knowledge

- *Introduction to waves (Unit 2 - Topic 5).*

- *Division of amplitude (Unit 2 - Topic 6).*

- *Reflection and Refraction of Light and Snell's Law.*

Learning objectives

By the end of this topic you should be able to:

- *understand the difference between plane polarised and unpolarised waves;*

- *to understand Malus' Law of Polarisation and Brewster's Law of Polarisation;*

- *to be able to derive the equation for Brewster's angle $n = \tan i_p$;*

- *to be able to explain some applications for polarisation.*

9.1   Introduction

One of the properties of a transverse wave is that it can be polarised. This means that all the oscillations of the wave are in the same plane. In this topic we will investigate the production and properties of polarised waves. Most of this topic will deal with light waves, and some of the applications of polarised light will be described at the end of the topic.

Light waves are electromagnetic waves, made up of orthogonal (perpendicular) oscillating electric and magnetic fields. When we talk about the oscillations of a light wave, we will be describing the oscillating electric field. For clarity, the magnetic fields will not be shown on any of the diagrams in this chapter - this is the normal practice when describing electromagnetic waves.

Top tip

Suggested practicals:

- Investigate polarisation of microwaves and light.

- Investigate reflected laser (polarised) light from a glass surface through a polarising filter as the angle of incidence is varied.

- Investigate reflected white light through a polarising filter.

9.2   Polarised and unpolarised waves

Let us consider a transverse wave travelling in the *x*-direction. Although we will be concentrating on light waves in this topic, it is useful to picture transverse waves travelling along a rope. Figure 9.1 shows transverse waves oscillating in the *y*-direction.

Figure 9.1: Transverse waves with oscillations in the y-direction

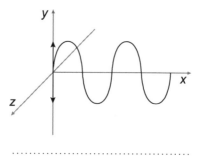

The oscillations are not constrained to the *y*-direction (the vertical plane). The wave can make horizontal oscillations in the *z*-direction, or at any angle ϕ in the *y-z* plane, so long as the oscillations are at right angles to the direction in which the wave is travelling (see Figure 9.2).

Figure 9.2: Transverse waves oscillating (a) in the z-direction, and (b) at an angle in the y-z plane.

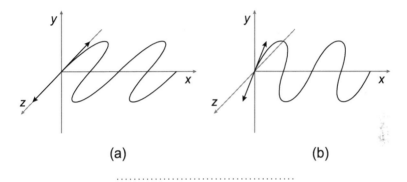

(a)　　　　　(b)

When all the oscillations occur in one plane, as shown in Figure 9.1 and Figure 9.2, the wave is said to be **polarised**. If oscillations are occurring in many or random directions, the wave is **unpolarised**. The difference between polarised and unpolarised waves is shown in Figure 9.3.

Figure 9.3: (a) Polarised, and (b) unpolarised waves

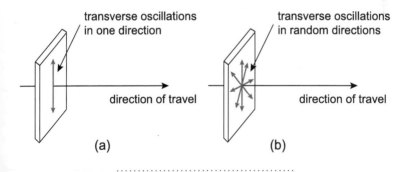

Light waves produced by a filament bulb or strip light are **unpolarised**. In the next two Sections of this topic different methods of producing **polarised light** will be described. You should note that longitudinal waves cannot be polarised since the oscillations occur in the direction in which the wave is travelling. This means that sound waves, for example, cannot be **polarised**.

9.3 Polaroid and Malus' law

Before we look at how to polarise light waves, let us think again about a (polarised) transverse wave travelling along a rope in the *x*-direction with its transverse oscillations in the *y*-direction. In Figure 9.4 the rope passes through a board with a slit cut into it. Figure 9.4 (a) shows what happens if the slit is aligned parallel to the *y*-axis. The waves pass through, since the oscillations of the rope are parallel to the slit. In Figure 9.4 (b), the slit is aligned along the *z*-axis, perpendicular to the oscillations. As a result, the waves cannot be transmitted through the slit.

Figure 9.4: (a) Transverse waves passing through a slit parallel to its oscillations, and (b) a slit perpendicular to the oscillations blocking the transmission of the waves

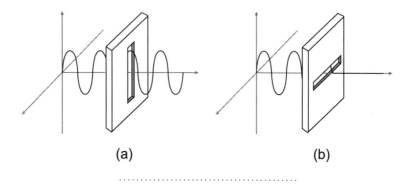

(a)   (b)

What would happen if the wave incident on the slit was oscillating in the *y*-*z* plane, making an angle of 45° with the *y*-axis, for instance? Under those conditions, the amplitude of the incident wave would need to be resolved into components parallel and perpendicular to the slit. The component parallel to the slit is transmitted, whilst the component perpendicular to the slit is blocked. The transmitted wave emerges polarised parallel to the slit. Later in this Section we will derive the equation used for calculating the irradiance of light transmitted through a polariser whose transmission axis is not parallel to the plane of polarisation of the light waves.

9.3.1 Polarisation of light waves using Polaroid

Although the mechanical analogy is helpful it cannot be carried over directly to the comparable situation involving light waves. Consider a sheet of Polaroid, a material consisting of long, thin polymer molecules (doped with iodine) that are aligned with each other. Because of the way a polarised light wave interacts with the molecules, the sheet of Polaroid only transmits the components of the light with the electric field vector perpendicular to the molecular alignment. The direction which passes the polarised light waves is called the **transmission axis**. The Polaroid sheet blocks the electric field component that is parallel to the molecular alignment.

Figure 9.5: Action of a sheet of Polaroid on unpolarised light

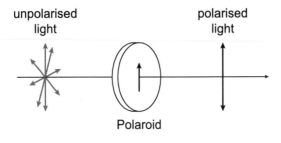

In Figure 9.5, light from a filament bulb is unpolarised. This light is incident on a sheet of Polaroid whose transmission axis is vertical. The beam that emerges on the right of the diagram is polarised in the same direction as the transmission axis of the Polaroid. Remember, this means that the electric field vector of the electromagnetic wave is oscillating in the direction shown.

Polarised light

There is an animation available online showing the electric and magnetic fields of a polarised light wave.

Go online

9.3.2 Malus' law

Earlier in this Section, the problem of a polariser acting on a polarised beam of light was introduced. We will now tackle this problem and calculate how much light is transmitted when the transmission axis of a Polaroid is at an angle to the plane of polarisation. The two cases illustrated in Figure 9.4 show what would happen if the transmission axis is parallel or perpendicular to the polarisation direction of the beam. In the former case all of the light is transmitted, in the latter case none of it is. Figure 9.6 shows what happens when the transmission axis of a sheet of Polaroid makes an angle θ with the plane of polarisation of an incident beam.

Figure 9.6: Polarised beam incident on a second sheet of Polaroid

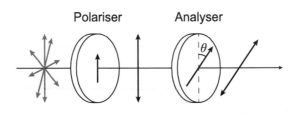

In Figure 9.6 an unpolarised beam of light is polarised by passing it through a sheet of Polaroid. The polarised beam is then passed through a second Polaroid sheet, often called the analyser. The transmission axis of the analyser makes an angle θ with the plane of polarisation of the incident beam. The beam that emerges from the analyser is polarised in the same direction as the transmission axis of the analyser.

Polariser and analyser

Go online

At this stage there is an online activity which shows the passage of a polarised light beam through two polarisers.

. . . .

A 'head-on' view of the analyser will help us to find the irradiance of the transmitted beam (see Figure 9.7).

Figure 9.7: Vertically polarised beam incident on an analyser

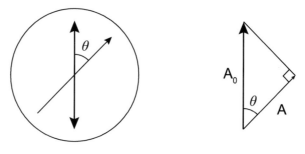

. . . .

The incident beam has amplitude A_0. From Figure 9.7, the component of A_0 parallel to the transmission axis of the analyser is $A_0\cos\theta$. So the beam transmitted through the analyser has amplitude A, where

$$A = A_0\cos\theta$$

(9.1)

. . . .

The irradiance of a beam, measured in W m^{-2}, is proportional to the square of the amplitude. Thus the irradiance I_0 of the incident beam is proportional to $A_0{}^2$ and the irradiance I of the transmitted beam is proportional to $A^2 = \left(= (A_0\cos\theta)^2 \right)$. From Equation 9.1

$$A = A_0 \cos \theta$$
$$\therefore A^2 = (A_0 \cos \theta)^2$$
$$\therefore A^2 = A_0{}^2 \cos^2 \theta$$
$$\therefore I = I_0 \cos^2 \theta$$

(9.2)

. . . .

Equation 9.2 is known as **Malus' law**, and gives the irradiance of the beam transmitted through the analyser.

Example

A sheet of Polaroid is being used to reduce the irradiance of a beam of polarised light. What angle should the transmission axis of the Polaroid make with the plane of polarisation of the beam in order to reduce the irradiance of the beam by 50%?

We will use Malus' law to solve this problem, with I_0 as the irradiance of the incident beam and $I_0/2$ as the irradiance of the transmitted beam. Equation 9.2 then becomes

$$\frac{I_0}{2} = I_0 \cos^2 \theta$$
$$\therefore \cos^2 \theta = \frac{1}{2}$$
$$\therefore \cos \theta = \sqrt{\frac{1}{2}}$$
$$\therefore \theta = 45°$$

. . . .

Quiz: Polarisation and Malus' law

Q1: Light waves can be polarised. This provides evidence that light waves are

Go online

a) coherent.
b) stationary waves.
c) monochromatic.
d) longitudinal waves.
e) transverse waves.

. . . .

Q2: Unpolarised light passes through a sheet of Polaroid whose transmission axis is parallel to the y-axis. It then passes through a second Polaroid whose transmission axis is at 20° to the y-axis. At what angle is the plane of polarisation of the emergent beam?

a) Parallel to the y-axis.
b) At 10° to the y-axis.
c) At 20° to the y-axis.
d) At 70° to the y-axis.
e) Perpendicular to the y-axis.

. . . .

Q3: Can sound waves be polarised?

a) Yes, any wave can be polarised.
b) No, because the oscillations are parallel to the direction of travel.
c) Yes, because the oscillations are perpendicular to the direction of travel.
d) No, because sound waves are not coherent.
e) Yes, because they are periodic waves.

. . . .

Q4: A polarised beam of light is incident on a sheet of Polaroid. The angle between the plane of polarisation and the transmission axis is 30°. If the irradiance of the incident beam is 8.0×10^{-4} W m^{-2}, what is the irradiance of the transmitted beam?

a) 2.0×10^{-4} W m^{-2}
b) 4.0×10^{-4} W m^{-2}
c) 6.0×10^{-4} W m^{-2}
d) 6.9×10^{-4} W m^{-2}
e) 8.0×10^{-4} W m^{-2}

. . . .

Q5: A polarised beam of light of irradiance 3.00 mW m^{-2} is incident on a sheet of Polaroid. What is the angle of the transmission axis relative to the incident beam's plane of polarisation if the transmitted beam has irradiance 1.00 mW m^{-2}?

a) 8.40°
b) 30.0°
c) 35.3°
d) 54.7°
e) 70.5°

. . . .

9.4   Polarisation by reflection

Light reflected by the surface of an electrical insulator is partially, and sometimes fully, polarised. The degree of polarisation is determined by the angle of incidence of the beam and the refractive index of the reflecting material.

Figure 9.8: Polarisation by reflection

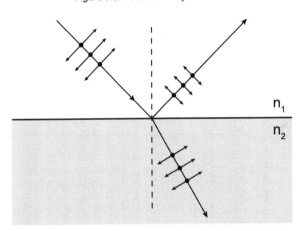

In Figure 9.8 the solid circles represent the components of the incident beam that are polarised parallel to the surface of the reflecting material. The double-headed arrows represent the components at right angles to those shown by the circles. The refracted (transmitted) beam contains both of these components, although the component in the plane of incidence is reduced. The refracted beam is therefore partially polarised, but the reflected beam can be completely polarised parallel to the reflecting surface and perpendicular to the direction in which the beam is travelling.

Usually the reflected beam is not completely polarised, and contains some of the 'arrows' components. We shall look now at the special case in which the reflected beam does become completely polarised.

9.4.1   Brewster's law

The Scottish physicist Sir David Brewster discovered that for a certain angle of incidence, monochromatic light was 100% polarised upon reflection. The refracted beam was partially polarised, but the reflected beam was completely polarised parallel to the reflecting surface. Furthermore, he noticed that at this angle of incidence, the reflected and refracted beams were perpendicular, as shown in Figure 9.9.

Figure 9.9: Brewster's law

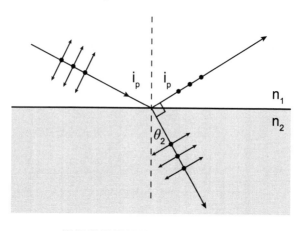

Light is travelling in a medium with refractive index n_1, and being partially reflected at the boundary with a medium of refractive index n_2. The angles of incidence and reflection are i_p, the polarising angle. The angle of refraction is θ_2. Snell's law for the incident and refracted beams is $n_1 \sin i_p = n_2 \sin \theta_2$.

According to Brewster

$$i_p + \theta_2 = 90°$$
$$\therefore \theta_2 = 90° - i_p$$

We can substitute for $\sin \theta_2$ in the Snell's law equation

$$n_1 \sin i_p = n_2 \sin \theta_2$$
$$\therefore n_1 \sin i_p = n_2 \sin (90 - i_p)$$
$$\therefore n_1 \sin i_p = n_2 \cos i_p$$
$$\therefore \frac{\sin i_p}{\cos i_p} = \frac{n_2}{n_1}$$
$$\therefore \tan i_p = \frac{n_2}{n_1}$$

(9.3)

....

This equation is known as **Brewster's law**. Usually the incident beam is travelling in air, so $n_1 \approx 1.00$, and the equation becomes $\tan i_p = n_2$. The polarising angle is sometimes referred to as the Brewster angle of the material.

Example

What is the polarising angle for a beam of light travelling in air when it is reflected by a pool of water ($n = 1.33$)?

Using Brewster's law

$$\tan i_p = n_2$$
$$\therefore \tan i_p = 1.33$$
$$\therefore i_p = 53.1°$$

....

The refractive index of a material varies slightly with the wavelength of incident light. The polarising angle therefore also depends on wavelength, so a beam of white light does not have a unique polarising angle.

Brewster's law

At this stage there is an online activity which allow you to investigate polarisation by reflection.

Go online

....

9.5 Applications of polarisation

9.5.1 Other methods of producing polarised light

Two methods for polarising a beam of light have been discussed. You should be aware that other techniques can be used. **Birefringent** materials such as calcite (calcium carbonate) have different refractive indices for perpendicular polarisation components. An unpolarised beam incident on a calcite crystal will be split into two beams polarised at right angles to each other. **Dichroic** crystals act in the same way as Polaroid. Their crystal structure allows only light with electric field components parallel to the crystal axis to be transmitted.

9.5.2 Applications of polarisation

Polarised light can be used to measure strain in **photoelastic** materials, such as glass and celluloid. These are materials that become birefringent when placed under mechanical stress. One application of this effect is in stress analysis. A celluloid model of a machine part, for example, is placed between a crossed polariser and analyser. The model is then placed under stress to simulate working conditions. Bright and dark fringes appear, with the fringe concentration highest where the stress is greatest. This sort of analysis gives important information in the design of mechanical parts and structures.

Figure 9.10: Shows this technique used with plastic forks

Polaroid sunglasses and camera lens filters are often used to reduce glare. In the ideal case, light reflected from a horizontal surface will be polarised in a horizontal plane as described earlier, so a Polaroid with a vertical transmission axis should prevent transmission completely. In practice, reflected light is only partially polarised, and is not always being reflected from a horizontal surface, so glare is only partially reduced by Polaroid sunglasses and filters.

Optically active materials can change the plane of polarisation of a beam of light. This process comes about because of the molecular structure of these materials, and has been observed in crystalline materials such as quartz and organic (liquid) compounds such as sugar solutions. The degree of optical activity can be used to help determine the molecular structure of these compounds. In the technique known as **saccharimetry**, the angle of rotation of the plane of polarisation is used as a measure of the concentration of a sugar solution. Polarised light is passed through an empty tube, and an analyser on the other side of the tube is adjusted until no light is transmitted through it. The tube is then filled with the solution, and the analyser is adjusted until the transmission through it is again zero. The adjustment needed to return to zero transmission is the angle of rotation.

Perhaps the most common everyday use of optical activity is in **liquid crystal displays** (LCDs). A typical LCD on a digital watch or electronic calculator consists of a small cell of aligned crystals sandwiched between two transparent plates between a crossed polariser and analyser. This arrangement is shown schematically in Figure 9.11.

Figure 9.11: LCD display with no field applied (a) and with a field applied (b)

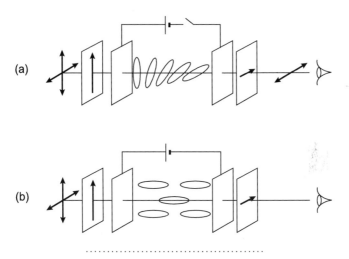

When no electric field is applied across the cell, the liquid crystal molecules are arranged in a helical twist. The polarised light entering the cell has its polarisation angle changed as it travels through the cell, and emerges polarised parallel to the transmission axis of the analyser. The cell appears light in colour, and is thus indistinguishable from the background. When a field is applied, the liquid crystal molecules align in the same direction (the direction of the electric field), and do not change the polarisation of the light. The emerging light remains polarised perpendicular to the analyser transmission axis. This light is not transmitted by the analyser, and so the cell looks dark. In this case, a black segment is seen against a lighter background.

Polarising filters are also used lots in photography, especially for intensifying the colour in images. Blue skies in many photos tend to look washed out and polarising filters are used to darken the sky and remove distracting reflections.

The lenses of some 3D polaroid glasses are both polarising lenses but they are aligned at right angles to each other, meaning that light which can pass through one is blocked by the other and vice versa. Similar to the older red/green or red/blue 3D glasses, they work by delivering slightly different views to each eye to produce a 3D floating image. This is commonly seen in 3D cinemas and has the major advantage that they allow full colour displays which the red/green ones always lacked.

The newest type of 3D glasses is the RealD 3D system which uses circularly polarised light instead of the linear polarisation described above. The advantage of this is that the person wearing the glasses can tilt their head without seeing double or darkened images.

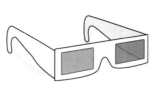

Quiz: Brewster's law and applications of polarisation

Go online

Q6: What is the polarisation angle for monochromatic light travelling in air, incident on a sheet of glass of refractive index 1.52?

a) 0.026°
b) 33.3°
c) 41.1°
d) 48.8°
e) 56.7°

....

Q7: Light reflected at the polarising angle is

a) polarised parallel to the reflecting surface.
b) polarised perpendicular to the reflecting surface.
c) polarised parallel and perpendicular to the reflecting surface.
d) polarised in the direction in which the wave is travelling.
e) completely unpolarised.

....

Q8: It is found that a beam of light is 100% polarised when reflected from a smooth plastic table-top. If the angle of incidence is 61.0°, what is the refractive index of the plastic?

a) 1.00
b) 1.14
c) 1.31
d) 1.80
e) 2.06

....

Q9: Which of the following sentences best describe a material that exhibits 'optical activity'?

a) The material changes its polarisation when under mechanical stress.
b) The material can rotate the plane of polarisation of a beam of light.
c) The material does not transmit polarised light.
d) The material reflects polarised light.
e) The material has different refractive indices for perpendicular polarisation components.

....

Q10: Polaroid sunglasses are most effective at reducing glare when the transmission axis of the Polaroid is

a) vertical.
b) horizontal.
c) at the polarising angle.
d) at 45° to the horizontal.
e) at 30° to the horizontal.

....

9.6   Extended information

Web links

There are web links available online exploring the subject further.

. . . .

9.7   Summary

Summary

You should now be able to:

- understand the difference between plane polarised and unpolarised waves;

- to understand Malus' Law of Polarisation and Brewster's Law of Polarisation;

- to be able to derive the equation for Brewster's angle $n = \tan i_p$;

- to be able to explain some applications for polarisation.

9.8   Assessment

End of topic 9 test

Go online

The following test contains questions covering the work from this topic.

Q11: An unpolarised beam of light is incident on a Polaroid filter. The emerging beam then travels through a second Polaroid filter. The first filter has its transmission axis at $22°$ to the vertical, the second is at $60°$ to the vertical.

State the angle of polarisation of the beam emerging from the second Polaroid, giving your answer in degrees to the vertical.

_ _ _ _ _ _ _ _ _ _ $°$

. . . .

Q12: A beam of polarised light of irradiance 4.3 W m^{-2} is incident on a sheet of Polaroid.

If the transmission axis of the Polaroid makes an angle $40°$ with the plane of polarisation of the incident beam, calculate the irradiance of the transmitted beam.

_ _ _ _ _ _ _ _ _ _ W m^{-2}

. . . .

Q13: Three Polaroid filters are lined up along the x-axis. The first has its transmission axis aligned in the y-direction, the second has its axis at $45°$ in the y - z plane, and the third has its axis in the z-direction.

1. Calculate the irradiance of the transmitted beam emerging from the third filter.

_____ W m^{-2}

2. In which direction is the emerging beam polarised?

a) In the z-direction.

b) At 45° in the y - z plane.

c) In the y-direction.

d) In the x-direction.

....

Q14: Calculate the angle, in degrees, required between the incident plane of polarisation and the transmission axis of a sheet of Polaroid to reduce the irradiance of a beam of polarised light from 7.2 W m^{-2} to 1.9 W m^{-2}.

_____ °

....

Q15: The polarising angle for a particular glass is 58.2°.

Calculate the refractive index of the glass.

_____

....

Q16: Calculate the polarising angle, in degrees, for a beam of light travelling in water ($n_w = 1.33$) incident on a block of leaded glass ($n_g = 1.54$).

_____ °

....

Q17: Light travelling in air is partially reflected from a glass block. The reflected light is found to be 100% polarised when the angle of incidence is 57.7°.

1. State the magnitude of the angle between the reflected and refracted beams.

_____ °

2. Calculate the refractive index of the block.

_____

....

Q18: A beam of light is 100% polarised by reflection from the surface of a swimming pool. The reflected beam is then passed through a Polaroid. The transmission axis of the Polaroid makes an angle of 38.5° with the normal to the plane of polarisation of the reflected beam.

If the reflected beam has irradiance 7.75 W m^{-2}, calculate the irradiance of the beam transmitted by the Polaroid.

_____ W m^{-2}

....

Topic 10

End of unit test

End of unit 2 test

Go online

DATA SHEET

Common Physical Quantities

The following data should be used when required:

Quantity	Symbol	Value
Gravitational acceleration on Earth	g	$9.8 \ m \ s^{-2}$
Mass of an electron	m_e	$9.11 \times 10^{-31} \ kg$
Planck's constant	h	$6.63 \times 10^{-34} \ J \ s$
Speed of light in a vacuum	c	$3.00 \times 10^8 \ m \ s^{-1}$
Speed of sound	v	$340 \ m \ s^{-1}$
Refractive index of air		1.00
Refractive index of water		1.33

Q1: An electron thought of as a particle is orbiting a Hydrogen with a speed of $1.45 \times 10^6 \ m \ s^{-1}$.

1. Calculate the de Broglie wavelength of the electron.

   _____ m

2. Calculate the angular momentum of the electron if it is found in the n = 3 orbit.

   _____ kg m^2 s^{-1}

. . . .

Q2: A proton in a particle accelerator is travelling with velocity $1.45 \times 10^5 \ m \ s^{-1}$.

Calculate the de Broglie wavelength of the proton.

_____ m

. . . .

Q3: Helical motion of particles causes impressive effects such as the Northern Lights. This is because: (choose all that apply)

1. The component of the velocity perpendicular to the magnetic field causes the particle to continue its forward motion.

2. The Northern Lights are often green due to the Oxygen particles that the charged particles collide with.

3. All charged particles from space are pulled to the Earth's surface by the strength of the Earth's magnetic field.

4. The component of the velocity perpendicular to the magnetic field causes circular motion.

5. The Northern lights are often blue due to the amount of Nitrogen in the atmosphere.

6. The component of the particle's velocity parallel to the magnetic field causes circular motion.

7. The component of the particle's velocity parallel to the magnetic field causes the particle to continue its forward motion.

....

Q4: An object of mass 0.15 kg is performing simple harmonic oscillations with amplitude 20 mm and periodic time 1.4 s.

1. Calculate the maximum value of the object's acceleration.

_____ m s^{-2}

2. Calculate the maximum value of the object's velocity.

_____ m s^{-1}

....

Q5: A travelling wave is represented by the equation:

$$y = 0.4 \sin \left(2\pi \left(3t - \frac{x}{1.7} \right) \right)$$

1. What is the value of the amplitude of this wave?

_____ m

2. Calculate the speed of the wave.

_____ m s^{-1}

....

Q6: A guitarist plays a note of 475 Hz on their guitar and at the same time a piano note is played at 473 Hz.

How many beats would be heard in 3 seconds of listening to the two notes?

_____ beats

....

Q7: Telecommunications signals are transmitted in fibre optic cable using infra-red light of wavelength (in air) 1.55×10^{-6} m. The fibre has a refractive index of 1.56.

1. Calculate the wavelength of the light in the fibre.

_____ m

2. For a light ray which travels straight down the fibre without making any reflections from its sides, calculate the optical path length of a 3.88 m section of fibre.

_____ m

....

Q8: The m = 10 dark fringe is observed for a wedge thickness 1.22×10^{-6} metres in an air-wedge interference experiment using monochromatic light.

At what wedge thickness is the m = 15 dark fringe observed?

_____ metres

....

Q9: The wavelength of a monochromatic light source is determined by a Young's slits experiment, in which the slits are separated by 5.45×10^{-4} m and placed 2.00 m from the viewing screen. The measured fringe spacing is 2.47×10^{-3} m.

Determine the value of the wavelength of the light obtained from this experiment.

_____ nm

. . . .

Q10: Light travelling in air is found to be 100% polarised when it is reflected from the surface of a pool of clear liquid at an angle of incidence $50.7°$.

1. Calculate the refractive index of the liquid.

   _____

2. In which direction is the reflected light polarised?

   a) Horizontally.

   b) In the same direction as the reflected ray is travelling.

   c) In the same direction as the incident ray is travelling.

   d) Vertically.

. . . .

Glossary

Amplitude

the maximum displacement of an oscillating object from the zero displacement (equilibrium) position.

Beat frequency

is the frequency of the changes between loud and quiet sounds heard by the listener - for example when a tuning fork and a piano note of similar frequencies are played at the same time.

Birefringence

the property of some materials to split an incident beam into two beams polarised at right angles to each other.

Black body

an object which absorbs all incident radiation and then re-emits all this energy again. Can be visualised as standing waves inside the blackbody cavity.

Bloomed lens

a lens that has been given a thin coating to make it anti-reflecting at certain wavelengths.

Bohr model

a model of the hydrogen atom, in which the electron orbits numbered $n = 1,2,3....$have angular momentum $\frac{nh}{2\pi}$.

Brewster's law

a beam of light travelling in a medium of refractive index n_1 will be 100% polarised by reflection from a medium of refractive index n_2 if the angle of incidence i_p obeys the relationship $\tan i_p = \frac{n_2}{n_1}$

Coherent waves

two or more waves that have the same frequency and wavelength, and similar amplitudes, and that have a constant phase relationship.

Compton scattering

the reduction in energy of a photon due to its collision with an electron. The scattering manifests itself in an increase in the photon's wavelength.

Damping

a decrease in the amplitude of oscillations due to the loss of energy from the oscillating system, for example the loss of energy due to work against friction.

De Broglie wavelength

a particle travelling with momentum p has a wavelength λ associated with it, the two quantities being linked by the relationship $\lambda = \frac{h}{p}$. λ is called the de Broglie wavelength.

Dichroism

the property of some materials to absorb light waves oscillating in one plane, but transmit light waves oscillating in the perpendicular plane.

Ferromagnetic

materials in which the magnetic fields of the atoms line up parallel to each other in regions known as magnetic domains.

Frequency

the rate of repetition of a single event, in this case the rate of oscillation. Frequency is measured in hertz (Hz), equivalent to s^{-1}.

Heliosphere

a massive bubble-like volume surrounded by the empty space.

Helix

the shape of a spring, with constant radius and stretched out in one dimension.

Interference by division of amplitude

a wave can be split into two or more individual waves, for example by partial reflection at the surface of a glass slide, producing two coherent waves. Interference by division of amplitude takes place when the two waves are recombined.

Interference by division of wavefront

all points along a wavefront are coherent. Interference by division of amplitude takes place when waves from two such points are superposed.

Irradiance

the rate at which energy is being transmitted per unit area, measured in $W\ m^{-2}$ or $J\ s^{-1}\ m^{-2}$.

Liquid crystal displays

a display unit in which electronically-controlled liquid crystals are enclosed between crossed polarisers. If no signal is supplied to the unit, it transmits light. When a signal is applied, the liquid crystal molecules take up a different alignment and the cell does not transmit light, hence appearing black.

Magnetic domains

regions in a ferromagnetic material where the atoms are aligned with their magnetic fields parallel to each other.

Magnetic poles

one way of describing the magnetic effect, especially with permanent magnets. There are two types of magnetic poles - north and south. Opposite poles attract, like poles repel.

Magnetosphere

the region near to an object such as a planet or a star where charged particles are affected by the object's magnetic field.

Malus' law

if a polarised beam of light, irradiance I_0, is incident on a sheet of Polaroid, with an angle θ between the plane of polarisation and the transmission axis, the transmitted beam has irradiance $I_0\cos^2$.

Optical activity

the effect of some materials of rotating the plane of polarisation of a beam of light.

Optical path difference

the optical path between two points is equal to the distance between the points multiplied by the refractive index. An optical path difference will exist between two rays travelling between two points along different paths if they travel through different distances or through media with different refractive indices.

Periodic time

the time taken to complete one cycle of the wave, measured in seconds.

Phase

a way of describing how far through a cycle a wave is.

Phase difference

if two waves that are overlapping at a point in space have their maximum and minimum values occurring at the same times, then they are in phase. If the maximum of one does not occur at the same time as the maximum of the other, there is a phase difference between them.

Photoelasticity

the effect of a material becoming birefringent when placed under a mechanical stress.

Photoelectric effect

the emission of electrons from a substance exposed to electromagnetic radiation.

Photoelectrons

an electron emitted by a substance due to the photoelectric effect.

Photons

a quantum of electromagnetic radiation, with energy $E = hf$, where f is the frequency of the radiation and h is Planck's constant.

Pitch

the distance travelled along the axis of a helix per revolution.

Polarisation

the alignment of all the oscillations of a transverse wave in one direction.

Polarised light

tight in which all the electric field oscillations are in the same direction.

Principle of superposition

this principle states that the total disturbance at a point due to the presence of two or more waves is equal to the algebraic sum of the disturbances that each of the individual waves would have produced.

Prominences

arcs of gas that erupt from the surface of the sun launching charged particles into space, sometimes called a filament.

Quantum mechanics

a system of mechanics, developed from quantum theory, used to explain the properties of particles, atoms and molecules.

Saccharimetry

a technique that uses the optical activity of a sugar solution to measure its concentration.

Simple harmonic motion (SHM)

motion in which an object oscillates around a fixed (equilibrium) position. The acceleration of the object is proportional to its displacement, and is always directed towards the equilibrium point.

Speed

the speed of a wave is the distance travelled by a wave per unit time, measured in $m\ s^{-1}$.

Stationary wave

a wave in which the points of zero and maximum displacement do not move through the medium (also called a standing wave).

Transmission axis

the transmission axis of a sheet of Polaroid is the direction in which transmitted light is polarised.

Travelling wave

a periodic disturbance in which energy is transferred from one place to another by the vibrations.

Unpolarised light

light in which the electric field oscillations occur in random directions.

Van Allen radiation belts

regions around the Earth, caused by the Earth's magnetic field, that trap charged particles. Discovered in 1958 after the Explorer 1 mission.

Wave function

a mathematical function used to determine the probability of finding a quantum mechanical particle within a certain region of space.

Wavelength

the distance between successive points of equal phase in a wave, measured in metres.

Wave-particle duality

the concept that, under certain conditions, waves can exhibit particle-like behaviour and particles can exhibit wave-like behaviour.

Work function

in terms of the photoelectric effect, the work function of a substance is the minimum photon energy which can cause an electron to be emitted by that substance.

Hints for activities

Topic 1: Introduction to quantum theory

Quiz: Atomic models

Hint 1: See the section titled The Bohr model of the hydrogen atom.

Hint 2: See the section titled The Bohr model of the hydrogen atom - in particular the relationship for angular momentum.

Hint 3: See the section titled The Bohr model of the hydrogen atom.

Hint 4: See the section titled Atomic spectra.

Topic 2: Wave particle duality

Quiz: Wave-particle duality of waves

Hint 1: This is a straight application of $E = hf$.

Hint 2: The minimum photon energy equals the work function.

Hint 3: See the section titled Compton scattering.

Hint 4: $E = hf$ and

$$p = \frac{h}{\lambda}$$

Hint 5: This is a straight application of

$$p = \frac{h}{\lambda}$$

Quiz: Wave-particle duality of particles

Hint 1: See the section titled Wave-particle duality of particles.

Hint 2: This is a straight application of

$$\lambda = \frac{h}{p}$$

Hint 3: This is a straight application of

$$\lambda = \frac{h}{p}$$

Hint 4: See the section titled Diffraction.

Hint 5: See the section titled The electron microscope.

Topic 3: Magnetic fields and particles from space

Quiz: The force on a moving charge

Hint 1: How does the charge on a proton compare with the charge on an electron?

Hint 2: This is a straight application of $F = q\,v\,B$.

Hint 3: What is the size of the magnetic force acting on a charged particle moving parallel to magnetic field lines?

Hint 4: What is the direction of a magnetic force relative to the direction of motion of a charged particle?

Hint 5: This is an application of

$$qvB = \frac{mv^2}{r}$$

Quiz: Motion of charged particles in a magnetic field

Hint 1: See the section titled Helical motion.

Hint 2: See the section titled Helical motion.

Hint 3: This is a straight application of

$$r = \frac{mv \sin \theta}{qB}$$

Hint 4: See section entitled The Solar Wind.

Hint 5: See the section Charged particles in the Earth's magnetic field.

Topic 4: Simple harmonic motion

Quiz: Defining SHM and equations of motion

Hint 1: See the section titled Defining SHM.

Hint 2: $\omega = 2\pi f$.

Hint 3: See the section titled Defining SHM.

Hint 4: See the relationships in the section titled Equations of motion in SHM. Calculate ω. Maximum acceleration occurs when displacement equals amplitude.

Hint 5: See the relationships in the section titled Equations of motion in SHM.

Quiz: Energy in SHM

Hint 1: See the relationships in the section titled Energy in SHM.

Hint 2: The total energy of the system is constant.

Hint 3: Consider the relationship $E_k = \frac{1}{2}m\omega^2(a^2 - y^2)$.

Hint 4: $PE = KE$ when half of the total energy is kinetic and half is potential - find the displacement when the kinetic energy is half its maximum value.

Hint 5: Consider the relationship $E_k = \frac{1}{2}m\omega^2(a^2 - y^2)$ with $y = \frac{1}{2}a$.

Quiz: SHM Systems

Hint 1: See the section titled Mass on a spring - vertical oscillations. First, find ω and hence find the period.

Hint 2: See the section titled Simple pendulum. First, find ω and hence find the frequency.

Hint 3: See the section titled Mass on a spring - vertical oscillations. First, find ω and hence find the spring constant.

Hint 4: See the section titled Simple pendulum.

Hint 5: See the section titled Simple pendulum - how does ω vary with l ?

Topic 5: Waves

Quiz: Properties of waves

Hint 1: See the section titled Definitions.

Hint 2: Use $v = f\lambda$ twice.

Hint 3: See the figure The electromagnetic spectrum in the section titled Definitions.

Hint 4: Use $v = f\lambda$.

Hint 5: See the section titled Definitions.

Quiz: Travelling waves

Hint 1: Substitute the values in the equation.

Hint 2: Compare the options with the general expression for a travelling wave

$$y = A \sin 2\pi(ft - \frac{x}{\lambda})$$

.

Hint 3: Substitute the values in the equation.

Hint 4: Compare the equation with the general expression for a travelling wave

$$y = A \, \sin \, 2\pi (ft - \frac{x}{\lambda})$$

Hint 5: First work out the values of f and λ by comparing the equation with the general expression for a travelling wave

$$y = A \, \sin \, 2\pi (ft - \frac{x}{\lambda})$$

Quiz: Superposition

Hint 1: The amplitude of the disturbance at this point is equal to the difference between amplitudes of the two waves.

Hint 2: Consider the phase of the waves arriving at the listener from the two loudspeakers.

Hint 3: Divide the wavelength into the distance.

Hint 4: See the section titled Fourier Series.

Hint 5: Use the Phase Angle Equation and be careful with the wavelength.

Quiz: Stationary waves

Hint 1: How does the frequency of the third harmonic compare with the fundamental frequency?

Hint 2: See the section titled Stationary waves.

Hint 3: How many wavelengths are there on the string when the fundamental note is played?

Hint 4: See the section titled Stationary waves.

Hint 5: To find out how to calculate the wavelength see the activity titled Longitudinal stationary waves. Then use $v = f\lambda$.

Hint 6: What is the difference between the frequencies? What is the Beat Frequency? Read the section on Beats for more information.

Topic 6: Interference

Quiz: Coherence and optical paths

Hint 1: See the section titled Coherence.

Hint 2: Divide the wavelength into the distance.

Hint 3: The optical path length in the glass is equal to *(actual path length in the glass x refractive index)*.

Hint 4: Be careful to work out the optical path **difference**.

Hint 5: The optical path **difference** is equal to 4λ; remember that the ray in air travelled the thickness of the plastic.

Topic 7: Division of amplitude
Quiz: Thin film interference

Hint 1: See the section titled Optical path difference.

Hint 2: See the section titled Thin film interference.

Hint 3: Find the values of λ that give constructive interference.

Hint 4: See the end of the section titled Thin film interference.

Hint 5: See the end of the section titled Thin film interference for the correct relationship.

Wedge fringes

Hint 1: Make sure you are measuring all the distances on the same scale - use m throughout.

Quiz: Wedge fringes

Hint 1: See the relationship derived in the section titled Wedge fringes.

Hint 2: Consider the expression for constructive interference derived in the section titled Wedge fringes.

Hint 3: See the relationship derived at the end of the section titled Wedge fringes.

Hint 4: See the relationship derived in the section titled Wedge fringes. Find the thickness of the air wedge for $m = 1$.

Hint 5: Consider the effect on the optical path and the condition for constructive interference.

Topic 8: Division of wavefront
Quiz: Young's slits

Hint 1: See the title of this topic.

Hint 2: Apply the relationship derived in the section titled Young's slit experiment.

Hint 3: Consider the relationship for Δx derived in the section titled Young's slit experiment.

Hint 4: This is a straight application of the relationship derived in the section titled Young's slit experiment.

Hint 5: Work out 5 times Δx.

Topic 9: Polarisation

Quiz: Polarisation and Malus' law

Hint 1: See the introduction to this topic.

Hint 2: The plane of polarisation of the emergent beam is the same as the transmission axis of the last polarising filter through which it travelled.

Hint 3: Only transverse waves can be polarised.

Hint 4: This is a straight application of Malus' Law.

Hint 5: This is a straight application of Malus' Law.

Quiz: Brewster's law and applications of polarisation

Hint 1: This is a straight application of Brewster's Law.

Hint 2: See the section titled Brewster's Law.

Hint 3: This is a straight application of Brewster's Law.

Hint 4: See the section titled Applications of polarisation.

Hint 5: See the section titled Applications of polarisation.

Answers to questions and activities

1 Introduction to quantum theory

Hydrogen line spectrum (page 8)

Q1: Photon is absorbed as the electron needs to gain energy to move up an energy level.

Q2: The larger difference in energy levels causes the photon to have more energy and hence a higher frequency due to $E = hf$. UV is higher frequency than visible light.

Quiz: Atomic models (page 9)

Q3: b) the electron's angular momentum is quantised.

Q4: d) 4.22×10^{-34} kg m^2 s^{-1}

Q5: e) $r = \frac{n\lambda}{2\pi}$

Q6: a) photons can only be emitted with specific energies.

End of topic 1 test (page 15)

Q7: 4.22×10^{-34} kg m^2 s^{-1}

Q8: 33.24×10^{-10} m

Q9: 9.79×10^{-11} m

Q10: 3.19×10^{-19} J

Q11: 6.24×10^{-7} m

Q12: A,C and E

Q13: 2.51×10^{-32} m

2 Wave particle duality

Quiz: Wave-particle duality of waves (page 23)

Q1: a) 3.88×10^{-19} J

Q2: b) 9.85×10^{14} Hz

Q3: d) the wavelength of the scattered photons increases.

Q4: a) The blue photons have greater photon energy and greater photon momentum.

Q5: c) 2.65×10^{-32} kg m s^{-1}

Quiz: Wave-particle duality of particles (page 29)

Q6: a) particles exhibit wave-like properties.

Q7: d) 1.14×10^{-10} m

Q8: c) 1.7×10^{-27} kg

Q9: b) particles can exhibit wave-like properties.

Q10: b) the electrons have a shorter wavelength than visible light.

End of topic 2 test (page 31)

Q11: $f = 5.66 \times 10^{14}$ Hz

Q12: $\lambda = 3.72 \times 10^{-6}$ m

Q13: $E_k = 1.89 \times 10^{-19}$ J

Q14: $p = 1.56 \times 10^{-27}$ kg m s^{-1}

Q15: $p = 2.21 \times 10^{-27}$ kg m s^{-1}

Q16: a) The proton

Q17: b) The electron

Q18: $p = 5.78 \times 10^{-24}$ kg m s^{-1}

Q19: $\lambda_e = 1.15 \times 10^{-10}$ m

Q20: Quantum tunnelling is when an incident **electron** is thought of as a **wave** and part of the wave crosses a **barrier** . Due to the **uncertainty** of the wave's position, this allows the electron to pass through the barrier. This is how nuclear **fusion** is possible in the sun and how modern **transistors** have become so efficient and tiny.

3 Magnetic fields and particles from space

Quiz: The force on a moving charge (page 42)

Q1: b) The particles experience the same magnitude of force but in opposite directions.

Q2: c) 2.4×10^{-13} N

Q3: e) all of them

Q4: a) constant speed.

Q5: a) 3.16×10^{6} m s^{-1}

Quiz: Motion of charged particles in a magnetic field (page 51)

Q6: a) (i) only

Q7: c) helical, with the axis in the magnetic field direction

Q8: e) 5.69×10^{-6} m

Q9: e) Van Allen belts

End of topic 3 test (page 53)

Q10: 2.97×10^{-16} N

Q11:

1. 7.41×10^{-15}
2. 20.1 mm

Q12: 3.97×10^{-8} seconds

Q13: 2

Q14:

1. 4.2×10^{-4} m
2. 2.2×10^{-3} m

Q15: B,D and E

Q16: d) plasma

4 Simple harmonic motion

Quiz: Defining SHM and equations of motion (page 64)

Q1: b) displacement.

Q2: a) 1.27 Hz

Q3: d) Acceleration

Q4: e) 98.7 ms^{-2}

Q5: c) 0.84 ms^{-1}

Energy in simple harmonic motion (page 66)

Expected answer

1. Note that the sum of kinetic and potential energies is constant, so that energy is conserved in the SHM system

$$PE + KE$$
$$= \tfrac{1}{2}m\omega^2 y^2 + \tfrac{1}{2}m\omega^2 \left(a^2 - y^2\right)$$
$$= \tfrac{1}{2}m\omega^2 a^2$$

The total energy is independent of the displacement y.

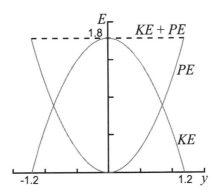

2. Again the sum of kinetic and potential energies is constant.

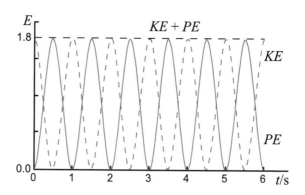

Quiz: Energy in SHM (page 67)

Q6: b) 0.037 J

Q7: d) 45 J

Q8: c) When its displacement from the rest position is zero.

Q9: c) $\pm a/\sqrt{2}$

Q10: d) 75 J

Quiz: SHM Systems (page 72)

Q11: d) 3.14 s

Q12: b) 0.910 Hz

Q13: e) 37 N m^{-1}

Q14: b) 0.75 s

Q15: a) Decrease its length by a factor of 4.

End of topic 4 test (page 75)

Q16: 1.0 m s^{-2}

Q17: 0.84 s

Q18:

   1. 3.1 m s^{-2}
   2. 0.86 m s^{-1}

Q19: 2.20 J

Q20: 0.25 m

Q21: 0.73 Hz

Q22: 1.5×10^{-3} J

Q23: 0.35 m

5 Waves

Quiz: Properties of waves (page 83)

Q1: e) amplitude

Q2: b) 4.3×10^{14} - 7.5×10^{14} Hz

Q3: a) X-rays, infrared, microwaves.

Q4: c) 4.74×10^5 GHz

Q5: d) 180 W m^{-2}

Quiz: Travelling waves (page 87)

Q6: c) 0.25 m

Q7: b) $y = 2\sin 2\pi \left(20t - x\right)$

Q8: e) 4 m

Q9: e) 12 Hz

Q10: b) 1.25 m s^{-1}

Quiz: Superposition (page 90)

Q11: d) 3.0 cm

Q12: a) a loud signal, owing to constructive interference?

Q13: c) 22.5 wavelengths

Q14: b) any periodic wave is a superposition of harmonic sine and cosine waves.

Q15: Phase angle = 3.7 rad

Superposition of two waves (page 91)

Q16: The resulting amplitude is equal to the sum of the amplitudes of A and B.

Q17: Once again the waves are back in phase, and so they interfere constructively again.

Q18: The amplitude is equal to the difference in the amplitudes of A and B, since the two waves are interfering destructively.

Longitudinal stationary waves (page 99)

Expected answer

a) (i)

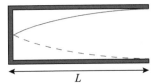

The fundamental has wavelength $4L$. Remember that λ is twice the distance between adjacent nodes, and so four times the distance between adjacent node and antinode.

a) (ii)

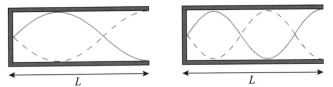

The wavelengths of the harmonics of a tube or pipe with one end open are $4L/3$, $4L/5$...

Note that only odd harmonics are possible for a tube with one end open.

b) (i)

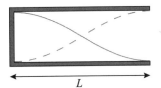

The fundamental has wavelength $2L$. Remember that λ is twice the distance between adjacent nodes or antinodes.

b) (ii)

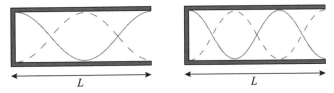

The wavelengths of the harmonics of a tube or pipe with both ends open are L, $2L/3$...

Quiz: Stationary waves (page 101)

Q19: c) 143 Hz

Q20: b) Every point between adjacent nodes of a stationary wave oscillates in phase.

Q21: e) 3.00 m

Q22: c) 4

Q23: d) 408 Hz

Q24: b) 25 beats

End of topic 5 test (page 103)

Q25: 5.54×10^{14} Hz

Q26: 6.52×10^{6} W m^{-2}

Q27: 272 cm

Q28: 8.8 m s^{-1}

Q29: 0.44 m

Q30: 3.1 cm

Q31: 0.61 m

Q32: 148 Hz

Q33: 0.24 m

Q34: 0.750 m

Q35: 26 m s^{-1}

Q36: 450 Hz

6 Interference

Quiz: Coherence and optical paths (page 112)

Q1: d) their phase difference is constant.

Q2: e) 9 wavelengths

Q3: c) 0.180 m

Q4: a) 8.75×10^{-3} m

Q5: b) 0.0682 m

End of topic 6 test (page 115)

Q6: 32.7 mm

Q7: 12.3 mm

7 Division of amplitude

Quiz: Thin film interference (page 123)

Q1:   c) their optical path difference is an integer number of wavelengths.

Q2:   b) 2.09×10^{-7} m

Q3:   c) 1120 nm and 373 nm

Q4:   d) The coating is only anti-reflecting for the green part of the visible spectrum.

Q5:   a) 9.06×10^{-8} m

Wedge fringes (page 128)

Expected answer

Use Equation 7.7, and make sure you have converted all the lengths into metres.

1. In the first case, the wavelength is 6.33×10^{-7} m.

$$\Delta x = \frac{\lambda l}{2y}$$
$$\therefore \Delta x = \frac{6.33 \times 10^{-7} \times 0.08}{2 \times 2 \times 10^{-5}}$$
$$\therefore \Delta x = \frac{5.064 \times 10^{-8}}{4 \times 10^{-5}}$$
$$\therefore \Delta x = 1.3 \times 10^{-3}\text{m} = 1.3\,\text{mm}$$

2. Now, the argon laser has wavelength 5.12×10^{-7} m.

$$\Delta x = \frac{\lambda l}{2y}$$
$$\therefore \Delta x = \frac{5.12 \times 10^{-7} \times 0.08}{2 \times 2 \times 10^{-5}}$$
$$\therefore \Delta x = \frac{4.096 \times 10^{-8}}{4 \times 10^{-5}}$$
$$\therefore \Delta x = 1.0 \times 10^{-3}\text{m} = 1.0\,\text{mm}$$

Quiz: Wedge fringes (page 128)

Q6:   d) 480 nm

Q7:   a) Blue

Q8:   d) 3.15×10^{-3} m

Q9:   b) 162 nm

Q10:   b) The fringes would move closer together.

End of topic 7 test (page 131)

Q11: 1.14×10^{-7} m

Q12:

1. 718 nm
2. 359 nm

Q13: 543 nm

Q14: 2.25×10^{-6} m

Q15: 4.21×10^{-3} m

Q16: 525 nm

8 Division of wavefront

Quiz: Young's slits (page 139)

Q1: b) interference by division of wavefront.

Q2: d) 333 nm

Q3: b) The fringes move closer together.

Q4: a) 2.4 mm

Q5: e) 0.016 m

End of topic 8 test (page 141)

Q6: 1.21×10^{-3} m

Q7: 491 nm

Q8: 441 nm

Q9: 2.23 mm

Q10: 535 nm

9 Polarisation

Quiz: Polarisation and Malus' law (page 149)

Q1: e) transverse waves.

Q2: c) At $20°$ to the y-axis.

Q3: b) No, because the oscillations are parallel to the direction of travel.

Q4: c) 6.0×10^{-4} W m^{-2}

Q5: d) $54.7°$

Quiz: Brewster's law and applications of polarisation (page 157)

Q6: e) $56.7°$

Q7: a) polarised parallel to the reflecting surface.

Q8: d) 1.80

Q9: b) The material can rotate the plane of polarisation of a beam of light.

Q10: a) vertical.

End of topic 9 test (page 158)

Q11: $60°$

Q12: 2.5 W m^{-2}

Q13:

   1. 2.1 W m^{-2}
   2. a) In the z-direction.

Q14: $59°$

Q15: 1.61

Q16: $49.2°$

Q17:

   1. $90°$
   2. 1.58

Q18: 3.00 W m^{-2}

10 End of unit test

End of unit 2 test (page 162)

Q1:

1. 5.02×10^{-10} m
2. 3.17×10^{-34} kg m^2 s^{-1}

Q2: 2.74×10^{-12} m

Q3: 2, 4 and 7 are the correct answers

Q4:

1. 0.40 m s^{-2}
2. 0.090 m s^{-1}

Q5:

1. 0.4 m
2. 5.1 m s^{-1}

Q6: 6 beats

Q7:

1. 9.94×10^{-7} m
2. 6.05 m

Q8: 1.83×10^{-6} metres

Q9: 673 nm

Q10:

1. 1.22
2. a) Horizontally.